GALILEO: PIONEER SCIENTIST

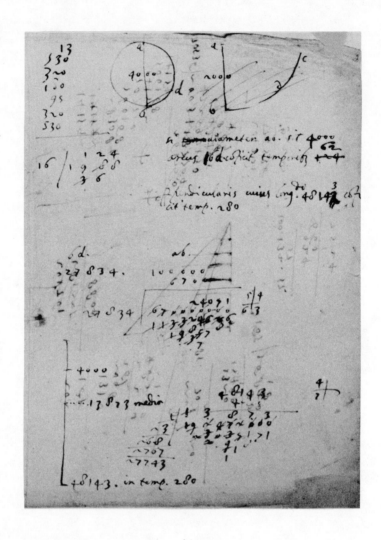

The page of Galileo's working papers on which he was writing when he recognized the times-squared law of distances in fall; f. 189v1 of vol. 72, mss. Galileiani. Reproduced with permission of the Biblioteca Nazionale Centrale, Florence. For English version see p. 20.

GALILEO: PIONEER SCIENTIST

Stillman Drake

UNIVERSITY OF TORONTO PRESS
Toronto Buffalo London

© University of Toronto Press 1990
Toronto Buffalo London
Printed in Canada

ISBN 0-8020-2725-3

Printed on acid-free paper

Canadian Cataloguing in Publication Data

Drake, Stillman
Galileo : pioneer scientist

Includes bibliographical references.
ISBN 0-8020-2725-3

1. Galilei, Galileo, 1564–1642. 2. Science, Renaissance.
3. Astronomers – Italy – Biography.
4. Physicists – Italy – Biography. I. Title.

QB36.G2D7 1990   520'.92   C89-095368-6

This book has been published with the help of a grant
from the Canadian Federation for the Humanities,
using funds provided by the Social Sciences and Humanities
Research Council of Canada. Publication has also been assisted by a
generous grant from the University of Toronto Women's Association.

To President Claude Bissell and Dean Ernest Sirluck
of the University of Toronto
and to the memory of Founding Director John W. Abrams
of its Institute for History and Philosophy of Science
this book is appreciatively dedicated.

Contents

Figures

Preface

This book integrates the scattered results of my researches since 1972, when details of Galileo's experimental work began to emerge from his working papers. Early results were included in my *Galileo at Work* of 1978. One problem, of determining step by step how Galileo discovered the times-squared law of distances in fall, was not resolved until 1986. Identification and dating of Galileo's steps from Ptolemaic astronomy to full Copernicanism required further study of manuscripts, particularly those of his youthful writing on motion and mechanics. Re-examination of the famous 1632 *Dialogue* then revealed reasons for its seemingly poor organization, as also for its inclusion of more physics but less astronomy than appeared to be promised by the (misleading) title under which that book has been reprinted and translated during the past two hundred years. Finally, Galileo's *Two New Sciences*, his last and scientifically his most important work, was perceived to contain an autobiographical passage corroborating all conclusions drawn from his working notes relating to the roles of the pendulum law and of inclined planes in his discovery of the law of fall. These (and other things) were still unknown when *Galileo at Work* was published; they came to light by completing the chronological arrangement of all surviving documents commenced in that book.

The further developments make it possible now to exhibit the entire career of Galileo in a new light, removing several puzzles that accumulated under traditional preconceptions and conjectures. How extensive the ramifications are will be evident to anyone who compares the following chronology of noteworthy events in Galileo's life to that supplied three decades ago in my *Discoveries and Opinions of Galileo*.

1564	Born at (or near) Pisa 15 February.
1574	Moves to Florence.
1575–80	Educated at Vallombrosa and Florence.
1581	Enters University of Pisa.
1583	Begins study of Euclid (outside the university).
1584	Composes *De universo*.
1585	Leaves University of Pisa (without a degree); teaches mathematics privately at Florence; constructs *bilancetta* and records specific gravities.
1586	Composes first scientific essay, *La bilancetta*; teaches publicly at Siena; begins writing a dialogue on problems of motion.
1587	Returns to private teaching at Florence; begins accumulating memoranda on motion; writes on centers of gravity of parabolic conoids; submits new theorem to Christopher Clavius at Rome.
1588	Shows the same theorem to Giuseppe Moletti at Padua; applies for chair of mathematics at University of Bologna; transmits his theorem to Guidobaldo del Monte at Pesaro; writes *Change and Elements*; composes Florentine *De motu*; lectures on Dante's *Inferno* to the Florentine Academy.
1589	Composes *Demonstration in Science* (on logic); becomes professor of mathematics at University of Pisa; begins revision of Florentine *De motu*; studies the *Almagest* and *De revolutionibus*.
1590	Conceives a geo-heliocentric (Tychonian) astronomy; begins composing Pisan *De motu*; writes geo-heliocentric commentaries on the *Almagest*.
1591	Adds inclined-plane and circular motions to *De motu*; conceives axial rotation of Earth (semi-Copernican); begins revision of his commentaries on the *Almagest*; death of Galileo's father, Vincenzio Galilei; termination of professorial contract at Pisa.
1592	Succeeds Moletti as mathematician at University of

Padua; learns of the astronomies of Tycho Brahe and of Ursus; writes on fortification and military architecture.

1593   Composes syllabus on mechanics.

1594   Expands syllabus on mechanics; composes *Cosmografia*.

1595   Explains tides in terms of Copernican motions of Earth; becomes fully Copernican in astronomy.

1596   Composes *Measurement by Sightings*; devises gunnery instrument based on Tartaglia's *squadra*.

1597   Defends Copernicus in letter to Mazzoni; avows his Copernicanism in letter to Kepler; devises first form of his military compass (the sector).

1598   Lectures on the pseudo-Aristotle *Problems of Mechanics*; revises the design of his geometric and military compass; composes manual for use of the compass.

1599   Finalizes form of geometric and military compass; employs an artisan to make mathematical instruments.

1600   Receives invitational letter from Tycho Brahe; begins composition of *De systemate mundi*; birth of eldest daughter, Virginia (Maria Celeste).

1601   Completes *De systemate mundi*; revises and greatly expands his treatise on mechanics; begins work on wheel of Aristotle and material strength; applies ratios of *moments* to Keplerian planetary data; birth of second daughter, Livia (Arcangela).

1602   Begins studies of magnetic phenomena; begins studies of long pendulums and motions on inclines; discovers 'Galileo's theorem' on times along chords; communicates findings on motions to Guidobaldo.

1603   Finds two least-time theorems for motions on inclines; afflicted with rheumatic illness, recurring thereafter; recognizes continued acceleration in natural motions; notes paradox of swiftnesses in accelerated motions; seeks rule of increase of speed during natural motions.

1604   Finds the odd-number rule of speeds by equalizing

	times; discovers the law of pendulums from careful timings; finds the law of fall as implied by the pendulum law; composes first attempted derivation of the law of fall; observes and lectures on the new star of October 1604.
1605	Publishes pseudonymous rustic dialogue on the new star; instructs Prince Cosimo de' Medici in mathematics; ridicules Copernicans in reprinted dialogue on new star.
1606	Prints manual for the geometric and military compass; birth of Vincenzio, his only son.
1607	Publishes account of plagiarism of his compass-manual; adds to his theorems relating to the law of fall.
1608	Resolves the paradox of swiftnesses in acceleration; discovers the parabolic trajectory by measurements; states the proportionality of speeds to times in fall.
1609	Prince Cosimo becomes Grand Duke of Tuscany; begins composition of treatise on natural motions; reports of the Dutch telescope confirmed to Galileo; Galileo's 9× telescope presented to Doge of Venice.
1610	Discovers four satellites of Jupiter ('Medicean stars'); publishes *Sidereus Nuncius*; resigns chair of mathematics at Padua; observes rings of Saturn, not resolved by his telescope; becomes chief mathematician and philosopher to Cosimo; observes phases of Venus, restoring his Copernicanism.
1611	Observes sunspots; visits Rome and exhibits his telescopic discoveries; compiles tables of motions for Jupiter's satellites; elected member of the Lincean Academy at Rome; debates with philosophers at Florence over floating bodies; Maffeo Barberini supports Galileo against philosophers.
1612	Publishes *Bodies in Water*; Tuscan philosophers form league to oppose Galileo; devises a micrometric device for his telescope; discovers eclipses of Jupiter's satellites; devises plan for determining longitudes at sea; Copernicus alleged by a priest to contradict the

Bible; writes against sunspot theory of Christopher Scheiner; records an observation of Neptune as a 'fixed star.'

1613   Galileo's *Letters on Sunspots* published at Rome; composes replies to four books against *Bodies in Water*; philosopher declares motion of Earth contradicts Bible; writes *Letter to Castelli* on science and religion.

1614   Galileo's daughters placed in convent at Arcetri; priest denounces Galileists from pulpit at Florence.

1615   *Letter to Castelli* sent by priest to Inquisition; Castelli replies to opponents of *Bodies in Water*; theologian publishes defense of Copernican astronomy; composes *Letter to Christina* on religion and science; hears rumors that Rome will suppress Copernican books; visits Rome to urge no church action on Copernicanism.

1616   Composes treatise on tides for Cardinal Orsini; edict regulating Copernican books issued at Rome; negotiates plan for longitude determinations with Spain.

1617   Improves tables of satellite motions; devises 'jovilabe' for calculating satellite positions.

1618   Resumes work on treatise on motion, laid aside in 1610; confined to bed by recurrent rheumatic condition; discusses with friends three comets near end of year.

1619   Grassi's book on comets misrepresents telescopic uses; composes *Discourse on Comets* (printed as Guiducci's); Grassi attacks Galileo as covertly a Copernican.

1620   Church 'corrections' to *De revolutionibus* published.

1621   Deaths of Grand Duke Cosimo II and Pope Paul V; devises doubly convex reading-glass.

1622   Composes *The Assayer* in reply to Grassi's attacks.

1623   *The Assayer* published at Rome by Lincean Academy; Maffeo Barberini is elected Pope Urban VIII.

1624   Visits Rome and is granted six papal audiences; proposes book using astronomical hypotheses in physics; composes *Reply to Ingoli* on theology and astronomy; begins writing *Dialogue on the Tides*.

1625   Complaint filed at Rome against *The Assayer*; Kepler criticizes both Galileo and Grassi.

1626   Corresponds with Cavalieri on motion and indivisibles; resumes work on magnetism and armaturing lodestones; studies theory of concave spherical mirrors.

1627   Resumes work on naturally accelerated motions; writes on theory of errors in estimation; serious recurrence of rheumatic illness.

1628   Corresponds with Cavalieri on paradoxes of infinites; recommends Cavalieri for chair at University of Bologna.

1629   Resumes composition of *Dialogue on the Tides*; finds sunspot evidence in support of Earth's motion.

1630   Visits Rome for licensing of *Dialogue on the Tides*; Pope requires that tides be removed from title of book; death of founder disbands Lincean Academy; requests licensing and printing of *Dialogue* at Florence; corresponds with Baliani on height of a column of water.

1631   Resumes work on motion and proof for the law of fall; writes on flood control for Bisenzio River; hostile book published by Jesuit Scheiner on sunspots.

1632   *Dialogue* published at Florence, 22 February; Urban VIII convenes panel to review complaints at Rome; pope turns matter of the *Dialogue* over to Inquisition; sale of *Dialogue* stopped and Galileo ordered to Rome.

1633   Arrives at Rome 13 February to face trial by Inquisition; interrogated 12 April under 'grave suspicion of heresy'; extrajudicial negotiation with prosecutor, 28 April; submits confession to lesser charge, 30 April; examined under threat of torture, 21 June; condemned to life imprisonment, 22 June; in custody of Archbishop of Siena, 9 July–15 December; at Siena, begins composing *Two New Sciences*.

1634	*Mechanics* translated into French by Marin Mersenne; composes *Postils to Rocco* defending his *Dialogue*.
1635	Learns that Inquisition forbids his printing any book; attempts to find printer in France or Germany.
1636	L. Elzevir agrees to print *Two New Sciences* in Holland.
1637	Completes *Two New Sciences*; composes *Astronomical Operations*; describes lunar librations.
1638	Becomes blind before receiving copy of *Two New Sciences*; dictates supplemental material, on force of percussion.
1639	Mersenne publishes French paraphrase of *Two New Sciences*.
1640	Dictates *Reply to Licetti*, requested by Prince Leopold.
1641	Declares no astronomical system to be worthy of belief; dictates design of a pendulum-escapement timekeeper; dictates critique of Euclid's Definition 5, Book v.
1642	Dies at Arcetri on 9 January.

Arranged in order, the events that have seemed too varied to be knit together closely appear as continuous, interrelated, and understandable. I shall put aside all opinions commonly debated among historians and philosophers of science about the sociology and philosophy best adapted to explain Galileo's career. Because scientists, who are likely to be interested in pioneering events, are not prone to accept anything unsupported by objective evidence, that will be emphasized instead. Where such emphasis is excessive for the purposes of other readers, it may safely be skipped over. The few footnotes pertain mainly to details of my researches. A bibliography is appended listing their piecemeal appearances, and the principal published attacks against them, together with books and articles relating to Galileo and the Scientific Revolution.

My principal source of documents is the National Edition of Galileo's works, edited by Antonio Favaro in twenty volumes over the years 1890–1910. Some working papers omitted from that are here presented when required, in facsimile reproductions made by kind permission of the Central National Library of Florence. Had all of these been published by Favaro, it is probable that physicists would long ago have arrived at the same analysis of experimental procedures employed by Galileo that will be offered here.

Once again I am happy to acknowledge much aid received from the University of Toronto, and from the John Simon Guggenheim Memorial Foundation, which made these researches possible. To James H. MacLachlan and Noel Swerdlow I am grateful for careful reading of the typescript and for corrections and suggestions of assistance to me. Above all, I once more thank Florence Selvin Drake for her patience and help during my years of concentration on the scientific career of the Italian pioneer of modern physics.

GALILEO: PIONEER SCIENTIST

Introduction

Einstein once advised us that in seeking to understand the
thought of a creative scientist, we should pay attention not just
to what he says, but to what he does. What Galileo did is known
mainly from surviving notes found on pages containing diagrams
and calculations but few or no words. What he said in books was
based upon what he had done, but was frequently misinterpreted
as long as we had only spotty information about the work that
lay behind his words. In astronomy, what Galileo had done was
told in his reports of telescopic observations and discoveries,
except for the measurements and calculations that lay behind
satellite astronomy. In the field of physics, Galileo's work
remained a subject of speculation. During the past fifteen years
the situation has greatly changed in both areas. We have now a
substantial amount of knowledge about Galileo's measurements
and calculations in physics, as well as in astronomy.

Measurements and calculations are of primary importance to
modern science, though they remain of little interest except to
scientists. Before Galileo, many volumes had been written on the
subject of motion (and on other parts of physics) by scholars who
had never actually measured any motions of falling bodies. The
making of measurements with great care and subjecting results
to mathematical analysis became principal activities of scientists
after the time of Galileo, and what they have said has been pretty
much limited to what they have learned from those activities. In
one sense that is why Einstein called attention to the dangers of
relying on what a scientist says without first having made sure

of what he does. Until the 1970s, historians of science writing about Galileo could only speculate from what he had said in his well-known writings, for lack of evidence about what he had done.

The most influential of those historians, the late Alexandre Koyré, concluded that Galileo had not *done* anything except with the telescope; for the rest (which included all his physics), he had only *thought*. Koyré did not reproach Galileo for that, but instead regarded it as Galileo's greatest contribution to science. To think mathematically about physical phenomena went against the grain of Aristotelian natural philosophy, which someone had to do before physics could be put on the right track. Koyré credited Galileo with placing physics on that road, and assigned to Plato full credit as the one who had inspired Galileo to do that.

This was in 1939, and for half a century now, students of Galileo have been writing mainly variations on a theme by Koyré. He believed he had proved that Galileo could not have measured actual motions accurately enough to have arrived at the law of falling bodies in that way. Galileo did not say he had; he stated only that in more than a hundred timings of balls rolling down inclined planes, none had departed from his law of natural descent by more than one-tenth of a pulse beat. Koyré declared any such published claim to be a fiction, seeing Galileo's description of his apparatus and procedures as entirely incapable of explaining his discoveries. How Galileo had *arrived* at his law of fall was not recounted explicitly in any of his published books, and that discovery was left still unexplained by Koyré.

Galileo arrived at the times-squared law of distances from rest in natural descent early in 1604 by first finding the law of the pendulum, relating that to fall, and fall to descents along planes. He left only one slender clue to his procedure in a final book, more than three decades after the events. But the whole story of discovery unfolds from Galileo's working papers. That story will be found in the first chapter of the present volume, for several reasons to be explained presently. Readers who are not versed in either physics or classical Euclidean mathematics may find it dull, but it will not lack interest for modern scientists. All of them have measured distances and times; hence they are fully aware of certain problems that arise when it becomes necessary to make the measurements with greater and greater precision. Those problems may differ according to the apparatus and procedures employed, but both the fact and the nature of such problems are inevitable, and were already present in the earliest actual measurements known to have been made of distances and times, using pendulums, descents along inclined planes, and vertical falls. How Galileo resolved problems of precise measurement as they arose is one part of the story that unfolds from his working papers on motion for the years 1604–9, and again from his journals and notes creating satellite astronomy during the years 1610–12.

The reason why few historians of science are currently interested in this story is that it can throw no light on Galileo as a philosopher. But as Galileo himself sarcastically inquired in 1605: 'What has philosophy got to do with measuring anything?' His reply, just one year after he had first succeeded in measuring times and distances in actual motions accurately enough to find the laws of pendulum and fall from his data, was that in matters of measurement it was the mathematicians and not any philosopher who should be trusted. For the past half-century historians of science have treated the word 'mathematicians' as synonymous with 'Platonists.' Hence Galileo's own reply would now be translated by them into: 'It is Plato that you have to trust above all other philosophers.' Koyré would approve; but that is not what Galileo wrote, and what he said may quite possibly be all that he meant.

Careful measurements of distances and times are made without conscious appeal to philosophy. The data yield information upon mathematical analysis – the same information to Aristotelians as to Platonists. How that information is interpreted may depend on philosophical preconceptions, unlike the processes of obtaining or of mathematically analyzing the data. The present book is largely concerned with the actual pioneer processes of this kind. Those belong to the history of physical science, and were legitimately employable according to the tenets of every philosophical school.

The history of the discovery of the law of falling bodies is not the only thing revealed by Galileo's working papers that has been overlooked or neglected in the past. The origin and gradual improvement of his tables of motions for four satellites of Jupiter deserve attention, for they opened a new branch of astronomy. In particular, the discovery of satellite eclipses gave Galileo, by new measurements, an argument for the Copernican system that he never made public. In this case what he did was not reflected in what he wrote, because his church had restricted what he could say in astronomy. Galileo neglected to recount his measurements in physics because they would not have interested contemporary readers, curious about nature rather than methods of discovery.

The purpose of this book is to inform everyone about the *pioneering* aspect of Galileo's thought. It is principally from the ordering and dating of his working papers and unpublished manuscripts, followed by re-examination of his printed books, that a coherent depiction emerges of him as a recognizably modern physical scientist whose pioneer investigation of gravitational phenomena has further potential applications even now.

The strictly chronological ordering that was adopted for my *Galileo at Work*, which was intended to provide historians with a convenient reference book of sequential activities, is unsuited to the portrayal of their relations to previous and to subsequent science, if not potentially misleading as to interrelations between sciences. Galileo's activities as a physicist altered the very nature of an ancient discipline (or pioneered a new one), whereas telescopic astronomy enriched an already established science – without changing its nature at all. The two crucial events were separated by only five years, but they were totally unrelated at the time. Galileo's interest in physics was continuous from the beginning of his career. His concern with astronomy was sporadic until the advent of the telescope – which indeed interrupted his work in physics for a considerable period. The most faithful portrait of Galileo as a scientist is one that shows him in the role of the pioneer modern physicist, and not in that of an over-zealous Copernican astronomer.

Accordingly I shall begin with discovery of the laws of the pendulum and fall, as marking the commencement of the early modern era in physics. After a chapter on the historical context of the law of fall, what it was that led Galileo to time motions carefully will next be recounted, together with his pre-modern astronomical reflections. The telescope later provided a new basis for his astronomy, not very different from his new basis for physics a bit earlier, for both were ultimately grounded in measurements.

The unit of length Galileo used in physics, which he called the *punto*, was 0.94 mm. It was quite arbitrary; he had a finely engraved brass rule divided into sixty equal parts, and had already recorded some careful measurements of distances before he began measuring times. At the start, his initial unit of time was also arbitrary – he weighed water that flowed at the rate of 3 fluid ounces per second and recorded those weights in grains, 1 ounce being 480 grains. But his final time unit, the *tempo* of 16 grains of flow, was not arbitrary, for it was related to his *punto* by the pendulum law. Galileo's *tempo* measured 1/92 second of time. In the final chapter it will be shown how the Galilean units still have potential utility.

One preliminary caution seems in order. Galileo did not use algebra and he never wrote an equation in his life, not even in his private papers. Algebra in equation form was just beginning to be adopted during his early years as professor of mathematics at Pisa. Since Galileo can hardly have failed to know that, his adherence to the Euclidean theory of ratios and proportionality among mathematically continuous magnitudes must be regarded as a deliberate choice on his part. Because very few persons today are likely to recall that theory, set forth in Euclid's *Elements*, Book v, Galileo's calculations may not be immediately clear to the reader, and may appear clumsy and inefficient. Nevertheless they were mathematically rigorous and correctly carried out.

Neither did Galileo use decimal fractions, which were first introduced in 1585. That was when Galileo began his scientific career, having left the University of Pisa (without a degree) in that year. He worked only in whole numbers and ratios of whole numbers. Historically regarded, neither algebraic equations nor decimal fractions had truly rigorous mathematical foundations until about a century ago, when Richard Dedekind redefined 'number' and devised the real-number system. The Euclidean theory of ratios and proportionality, in contrast, was rigorously founded in Greek antiquity, and Dedekind paid tribute to it as the ultimate basis underlying all correct treatments of number and of the continuum at his own time.

There is accordingly a sense in which Galileo's mathematical physics was more rigorous than were some practices in mathematics in use at his time by others more concerned with facility than with foundations. Those practices were eventually given mathematical rigor, but much later, and only by completely revising the basic concept of number. Meanwhile Galileo, and Newton after him, took care to avoid possible pitfalls by reasoning from ratios and proportionality (defined by Euclid as 'sameness of ratio') only, relations without commitment to numbers as absolutes. Leibniz, and Descartes before him, did not exercise the same caution, and thereby enhanced ease of calculation at the cost of mathematical rigor in procedure.

One consequence of Galileo's adherence to Euclid should be of particular interest to physicists. Our local gravitational constant g did not exist for Galileo. In one sense he did not recognize any physical constant, for physical constants arise in the writing of physical equations, and would simply cancel out in proportionalities. But we can factor one crucial calculation by Galileo, a calculation that put into his hands the times-squared law of distances in fall, in such a way as to disclose the ratio that served him as the equivalent of our g. It was the ratio of a distance fallen from rest to the length of pendulum timing that fall by swing to the vertical through a small arc. That ratio is simply $\pi^2/8$, a constant holding not only everywhere on earth, but also on the moon, or on Jupiter, or anywhere that bodies fall and pendulums oscillate (instead of just revolving uniformly forever).

How Galileo arrived at the *tempo* as his unit of time for the measurement of gravitational phenomena is the most interesting event in the whole history of the law of falling bodies, to which it is now appropriate to turn. Readers who prefer a strictly chronological approach, or who find chapter 1 perplexing, should postpone its reading until after chapter 6, or commence instead with the biographical chapter 3 or with background material that is to be found in chapter 2.

CHAPTER ONE

The Laws of
Pendulum and Fall

Galileo's working papers on motion from 1602 to 1637, when his *Two New Sciences* went to the publishers, are now bound into volume 72 of the Galilean manuscripts at Florence. On f. 107v of that volume are the first surviving measurements of distances that he recorded. They were in *punti*, though no unit was named on that page, which contained nothing except numbers and diagrams. A few days later the same measurements became linked with the law of fall, so let us begin with f. 107v. On that page are found measurements made with a ruler, probably brass, finely divided into 60 equal parts of 0.94 mm each. Galileo frequently used that same ruler when drawing diagrams in his working papers during the years 1604–5.

A grooved inclined plane slightly over 2100 *punti* long had been tilted 60 *punti*, to an angle of 1.7° with the horizontal. Along the groove a bronze ball descended from rest, repeatedly, while Galileo divided the time into 8 equal intervals, probably by singing a tune at beats of 0.55 second each. The place of the ball at each beat was marked; at those marks, strings were tied around the grooved plane. Probably gut strings were used, as when frets were tied around the neck of a lute – firmly, but capable of being adjusted when tuning the instrument. After the ball passed over each string, a faint bumping sound was audible as it struck the plane. Positions of the strings were patiently adjusted until all sounds coincided with notes of the tune, and the distance from resting contact of ball with plane to the lower side of each string was noted. Analysis shows that Galileo was accurate within

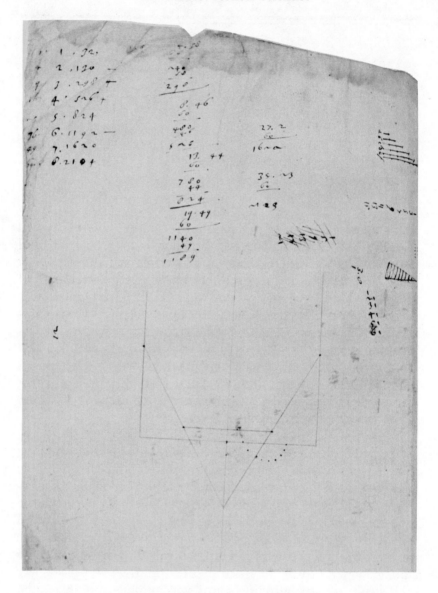

Figure 1. f. 107v. The square numbers in left margin confirmed
the law of fall to hold also for descent on an inclined plane.
The other entries had been made before the timing device
sketched at the end was constructed.

1/64 second for all strings except the lowest. Later he adjusted that string by about 20 *punti* (say 2 cm) and marked also a + or a – sign after four other original measures that seemed to Galileo a bit short or a bit long when he made subsequent reruns (these then sounding to him early, or late).

The purpose of this work had been to find, if possible, a rule for the increase of speed during natural straight descent of a heavy body. The Aristotelian word 'natural' was used for motions that took place spontaneously upon simple release, to distinguish those from 'forced' or 'violent' motions. Galileo found that during equal successive time intervals from rest, his measured speeds went up as do the odd numbers 1, 3, 5, 7, ... [1] This simple and exact rule doubtless surprised Galileo as much as it delighted him, for only a year before he had remarked in a letter that when put to a test with material objects, rules deduced by mathematicians simply did not agree with observation. Material impediments, as Galileo called friction, air resistance, absence of perfect smoothness and hardness, or the like, played a very important part in his physics at every stage from 1590 on, if not even earlier.

Having found a simple arithmetical law for a natural motion, Galileo thought next of *measuring* short intervals of time instead of merely equalizing them. Near the bottom of f. 107v he drew a diagram of the device that he used from then on for measuring the times of motions. In *Two New Sciences* he described its simplest form, in which flow of water from a large pail through a slender tube in the bottom was collected during each motion, was weighed on a sensitive balance, and the weights were treated as measures of the times. The sketch on f. 107v suggests (and the precision of Galileo's few surviving records of timings confirms) my belief that Galileo probably built and used a somewhat more reliable and consistent means of starting and ending flows than mere removal and replacement of the thumb or a finger at the

1 At this time Galileo could measure only overall speeds and did not yet have a way to *time* brief motions. Because the times had been equalized, the distances were direct measures of overall speeds. It did not occur to Galileo at first to consider them as simple *distances* also, so he did not sum the successive odd numbers and perceive the times-squared law at once.

end of the tube. That simple procedure has, however, been found to give results twice as reliable as Galileo asserted in his later book, and his own extant timings confirm this finding.[2]

The rate of flow of water from Galileo's device was 3 fluid ounces per second, very nearly indeed. This round number in an old standard unit was mere coincidence, because Galileo was not attempting to measure time in any accepted unit, let alone in astronomical seconds. He always worked entirely in ratios and proportionalities, as already mentioned, and *ratios* of times (or weights) are unit-free. He did indeed employ a standard unit of weight; namely, the grain, of which there were 480 to the fluid ounce. Galileo's recorded times are in grains weight of flow of water from his timing device, not in *tempi*. (One *tempo* would be measured by 16 grains weight of flow. By coincidence, 16 grains weight is almost exactly 1 gram in metric units.[3])

Galileo's first recorded timing was noted on f. 154v in the form of a column of numbers, totaled as $1000 + 107 + 107 + 107 + 16 = 1337$ (grains weight of water collected during a motion from rest). That was his timing of fall through 4000 *punti* = 376 cm. It was his least exact timing, being about 1/30 second too high. From his collection vessel Galileo first poured off 1000 grains, probably into a container marked as holding that amount. He then marked on the collection vessel the level at which the remainder stood, and later on he used that mark as representing 320 grains (the rounded sum of 3×107). The last figure, 16 (grains), was indirectly weighed; it represented the weight of the collection vessel damp less its dry weight. As a former medical

---

2 Galileo said that in more than a hundred timings along inclined planes, agreement with the times-squared law had been found never to vary more than one-tenth of a pulse beat. Dr Thomas B. Settle reported in 1961 that with little practice he had achieved about double that precision; cf. Settle, 'An experiment in the history of science,' *Science* 133 (1961), 19–23.

3 Settle avoided the nuisance of repeated weighings by collecting flows in a graduated cylinder and taking 1 cc as 1 gram. It is a curious fact that Galileo, though he did not have a graduated cylinder, used volumetric determinations in place of weighings by marking his collection vessel, as will be seen. That 1 gram of water represented 1 *tempo* for Galileo and was also the unit for Settle's ratios of times was purely coincidental, since the rates of flow were doubtless very different.

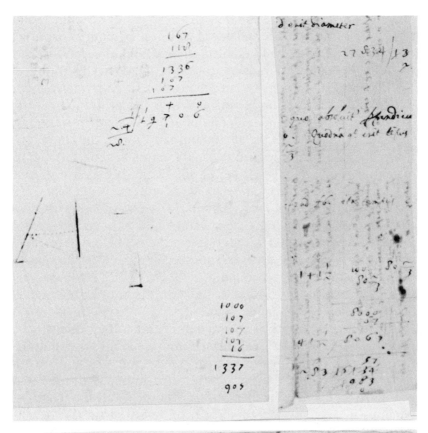

Figure 2.
f. 154v. First recorded timings of fall, and (later) a mean-proportional pendulum calculation.

f. 90bv. The surviving fragment of calculations used in discovery of the law of fall from the pendulum law. See text, pp. 18, 22, 26.

f. 115v. Correction of time of fall 4000 *punti*, using the new-found law.

student, Galileo was familiar with that way of accounting for fluid that adhered to the sides of a vessel after pouring from it.

Directly beneath 1337 Galileo entered 903, a second timing in grains flow during a motion from rest. That was his timing of fall from 2000 *punti* = 188 cm (about 6 feet), a very convenient height at which the timing with Galileo's device did not require difficult action on his part. (To dislodge a weight from double that height, as before, in exact coordination with the starting of flow, was less simple). Timing of 903 grains for fall 2000 *punti* is nearly exact; I calculate that 911 or 912 grains would be precise at the latitude of Padua (taking $g = 980.7$ cm/sec^2), and that Galileo's finding was little more than 1/200 second too low. In this case there is also evidence that Galileo had marked the level at which 903 grains stood in the collection vessel before he weighed the water. That is suggested by the fact that there was no column of figures, totaled, as before (and as on f. 189v1 a bit later), and it is confirmed by a figure on f. 151v that cannot be reasonably accounted for except as follows.

It should be remarked in explanation first that Galileo was concerned in this investigation with timing swings of pendulums through small arcs to the vertical, for reasons to be set forth presently. On f. 151v there are diagrams intended to represent geometrically some distance of fall from rest and the length of pendulum swinging to the vertical in a related time. There is also a freehand sketch of two meshed gears, and the calculation $53 \times 30 = 1590$. As mentioned before, Galileo measured with a carefully engraved ruler 60 *punti* long; also, he habitually first eliminated the fraction 1/2 before he made a multiplication or a division. The above calculation stands for $26\frac{1}{2} \times 60 = 1590$, the length in *punti* of a measured pendulum that, by my calculation, swung at Padua through a small arc to the vertical during flow of 903 grains of water through Galileo's timing device. The pendulum that takes the same time to the vertical as does fall through 2000, in any units whatever and anywhere, is in fact 1621 units long (to the nearest digit), and not 1590.

Thus Galileo's small error in timing fall through 2000 *punti* resulted in a shortage of 31 *punti* for the pendulum that would time that fall exactly by swing to the vertical through a small arc.

Although Galileo's two timings had been exactly consistent, his distance ratio, 2000/1590 = 1.2579, was considerably above the correct $\pi^2/8 = 1.2337$, or 2000/1621 to four significant places.

Galileo's skill as an experimentalist is illustrated by the pendulum length that he recorded on f. 151v as being 1590 *punti*. That figure can have been obtained only by finding the pendulum whose swing to the vertical through a small arc accompanied flow of 903 grains weight of water through his timing device. Implied is his having started with a pendulum about five feet long, and then having patiently adjusted it until water flowed precisely to the previously marked level while the pendulum swung to the vertical. This procedure throws light on the sketch of two meshed gears in the same working paper, seemingly quite out of place on a page of measurements of fall and pendulums. Galileo had not been working on mechanics after 1601; in 1602 he began on beam strength and on motion. Hence the sketch of gears in 1604 was puzzling until the full context of Galileo's experimental work on f. 151v became clear. When he had found the pendulum for 903 grains flow, it would be natural for Galileo to consider the pendulum for 1337 grains. It would be very long and hence inconvenient to alter in length repeatedly as before. Running the string over a nail in a movable upright and anchoring it to a bench would allow the nail to be raised and lowered by gears and a crank. Such a scheme accounts for the sketch. In the end, however, Galileo timed the pendulum for $1337/2 = 668\frac{1}{2}$ grains of flow as being 870 *punti* in length, and then he timed the doubled pendulum of 1740 *punti* at 942 grains of flow. At any rate that is what is implied by the working papers that Galileo preserved.

As mentioned earlier, Galileo timed pendulum swings to the vertical only for his pendulums, not the full period as now defined by a full swing and return. He used the quarter-period because the only swing he could time with precision was from the instant of releasing the bob to the sound of impact with a block that was fixed in advance against a side of the bob when hanging plumb. Ever since the time of Galileo's contemporary and critic, Marin Mersenne, historians have assumed that Galileo failed to notice

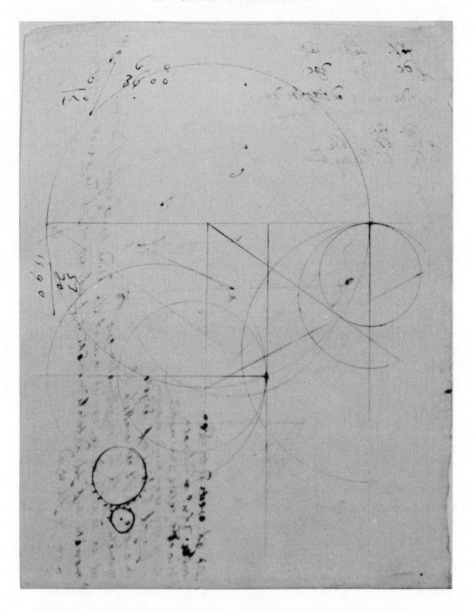

Figure 3. f. 151v. Galileo's notation of 1590 *punti* as the pendulum timing flow of 903 grains of water, on a page containing his geometric models for pendulum and fall.

that perceptible time differences exist between moderate and very wide swings of the same pendulum. Mersenne noted this fact by swinging two equal pendulums, released simultaneously, through different arcs to the same block. Some statements about pendulums in Galileo's last book caused Mersenne, one of the ablest living experimentalists, to doubt that Galileo had ever recognized this fact. Galileo's statements can now be better understood in the light of his unpublished papers here under consideration, and they will be taken up in chapter 12.

From the pendulum measurements Galileo now had at hand, a table of the kind shown below could be compiled; though I doubt that Galileo troubled to write a table, mine will serve to show the way in which he arrived at his discoveries by simply applying the theory of ratio and proportionality set forth in Euclid's *Elements*, Book v. My table has been extended far enough to show the source of a very important number found in two of Galileo's surviving working papers, on one of which he was writing when he first recognized the law of fall in times-squared form, from its mathematically equivalent mean-proportional form.

Length of pendulum in *punti*	Time to the vertical in grains flow
870	$668\frac{1}{2}$
1,740	942
3,480	1,337
6,960	1,884
13,920	2,674
27,840	3,768

The first column was formed by successive doubling; the second, by alternate doublings. Except for a slight discrepancy with the second timing, each column separately is accordingly in continued proportion. Galileo's new time unit, the *tempo*, came into being when he related the two columns horizontally, so to speak. His original measure of time in grains of flow from a particular device having been completely arbitrary, he was free to alter it in any ratio he pleased. Taking each time to be the mean proportional between 2 and the length of pendulum, the two columns become related line by line. The same numbers result also, almost exactly, from division of each time in grains by 16, so 16 grains of flow became 1 *tempo* – the new unit adopted as a result of this investigation.

Because division by 16 of the times in grains weight of flow did not *exactly* produce that mean-proportional relation between the two columns, Galileo made an adjustment that resulted in his change of 27,840, as shown above, to 27,834. That work was done on one of the working papers of which only a part survives. On the blank side he wrote, in 1609, a note on another topic, cut it out, and pasted it on f. 90. That was lifted at my request, and on the hidden side I saw the number 27,834, twice, with enough words to identify is as a 'diameter.' Galileo's diagram and calculations were thrown away with the part of the page cut off, but they had been finished before Galileo discovered the law of fall, because 27,834 played a crucial part in the calculation on f. 189v1, which put that discovery into Galileo's hands.

With his adoption of the *tempo* as the unit of time, Galileo had the pendulum law in a restricted form; that is, for any set of pendulums successively doubled in length. It would have been a difficult task to test it for successively tripled pendulums, let alone for any other integral multiples, and quite impossible to establish it experimentally in its complete generality. What Galileo did next is seen on f. 154v (on which he had entered his first two timings). He now calculated the mean proportional of 118 and 167 – the times to the vertical, in *tempi* for the two pendulum lengths 6960 and 13,920 *punti* – getting 140 (*tempi*). The mean proportional of those two lengths is 9843, so if his restricted pendulum law were perfectly general, a pendulum 9843 *punti* long would swing to the vertical in 140 *tempi*.

On the other side of the same page, f. 154r, Galileo wrote the note *filo br. 16* – 'the string is 16 *braccia* long.' From two lines drawn and labeled by Galileo at Padua, I measured one *braccio* as containing about 620 *punti*. At 615 *punti* per *braccia*, length of the pendulum would be 9840 *punti*, or about 30 feet. Such a pendulum could be hung from a window over the courtyard of the University of Padua and timed, protected from wind. At Padua it would reach the vertical through a small arc in 141 *tempi*, by my calculations. Thus Galileo was fully assured of the complete

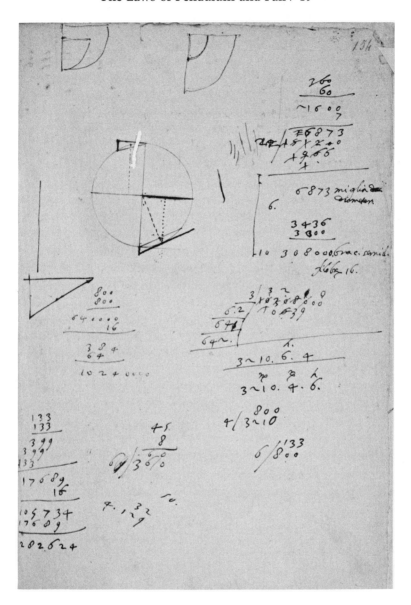

Figure 4. f. 154r. The note *filo br. 16*, at right center, was first entered. Having thus confirmed the general pendulum law, Galileo considered a pendulum as long as Earth's radius.

```
        13
       530
       320
       180
        95
       320
       580
 16)  1988  (124
```

If ~~semi~~diameter ab is 4000
arc bd is run in ~~124~~ 62 *tempi*

Vertical whose length is 48,143
is run in 280 *tempi*

bd	ab
27,834	100,000
	6,700

```
27,834)  670,000,000  (24,071 +
```

```
        48,143
         4,000
      192,572,000
```

4,000

13,873 mean [proportional]

48,143 in 280 *tempi*

Figure 5. English translation of f. 189v1; for original,
see frontispiece.

generality of his pendulum law in mean-proportional form that is mathematically the same as our law that periods of pendulums are as the square roots of their lengths.

In order to finish linking fall with the pendulum, Galileo needed one bit of information that has not yet been mentioned – the time of fall through a distance equal to the length of a timed pendulum. The length he chose was $2 \times 870 = 1740$ *punti*, and that fall takes 850 grains flow of water. The figure 850 exists in Galileo's surviving working papers only by implication. At top left on f. 189v1 there is a column of figures, added to total 1988, in which two separate pendulum timings are recognizable; I will subtotal its two parts in parentheses:

$$13 + 530 + 320 + 180 \ (= \ 1043) \text{ and } 95 + 320 + 530 \ (= \ 945),$$
$$\text{total } 1988$$

The numbers 180 and 95 represented actual weighings of collected water, while the number 13 represented adjustment for unweighed water adhering to the sides of the collecting vessel. The other two numbers, 530 and 320, were weights in grains found by pouring water from the collecting vessel to marks previously made. The mark for 320 grains had been made during the very first timing, as noted earlier. Hence the mark for 530 grains had been placed when a timing of 850 grains was made, and 850 grains of flow does time the fall through 1740 *punti*. Thus the column of figures totaled at the top of f. 189v1 recorded two separate timings of a pendulum of length 1740 *punti*, for reasons to be explained below. That pendulum was probably still hanging in place where Galileo had left it when he hung and timed the very long pendulum whose swing in 140 *tempi* had confirmed the general law of pendulums.

The work at the top of f. 189v1 is unrelated to the crucial calculation immediately below it, in which Galileo used the ratio $942/850 = 1.108$ to four places, of which the square is 1.228. Time to the vertical through a small arc, for any pendulum, has to time of fall through the distance equal to the length of the pendulum the ratio $\pi/2\sqrt{2} = 1.1107...$, so the above-measured time-ratio was quite creditable. In the crucial calculation on f. 189v1 it put into Galileo's hands the times-squared law of fall.

Discussion of that will be postponed until the unrelated work at the top of f. 189v1 has been explained.

When Galileo started writing f. 189, he intended to proceed with the adaptation of his pendulum law to distances and times of fall, but he became diverted into another series of reflections. In his preliminary calculations, now lost because of his later cutting and discarding part of f. 90^{b}v, on which he had obtained the number 27,834 as the 'diameter' of a pendulum, he found that in adapting the pendulum law to phenomena of fall he often had to double a length in deriving a time, and to double a doubled length in order to double a time. That suggested doubling the times measured to the vertical, and using time of swing from side to side, instead of time to the vertical only. Another diagram at the top of the page, later canceled, reflects this intention, as does the original wording of the first statement on f. 189v1, later halved to agree with time to the vertical only.

Contrary to common opinion, Galileo was already aware by the end of 1602 that a pendulum does not reach the vertical in exactly the same time through all arcs. When one arc is extremely wide, the swing takes perceptibly longer. But in 1602 Galileo had had no way of accurately timing swings of pendulums. In 1604 he proceeded to time two swings of the 1740-*punto* pendulum, first through a full quadrant (to judge from the canceled diagram on f. 189v1), and then only through a small arc. Having already timed the latter at 942 grains of flow, he questioned his timing of 1043 grains for the maximal arc, timed the small arc again, and got 945 grains. The excess time of his maximal swing over 942 was 9.3%, in good agreement with modern dynamic analysis for swings differing by 10° from minimal and maximal arcs.[4]

Twice in his later *Dialogue* (1632) Galileo had occasion to mention times of a pendulum through different arcs. At both places he said they were equal or insensibly different, which has

---

4 It is a great practical advantage for smoothness of swing to hold the bob a bit below the horizontal at the start, with some weight acting on the string at all times. In *Two New Sciences* Galileo gave 80°, where one would expect 90°, to indicate a very wide (or maximum practicable) swing.

appeared to confirm a belief that Galileo remained ignorant of the facts. Mersenne's technique certainly made the time difference 'sensible,' because the two sounds when bobs strike a block at the vertical are distinct and show that swing through the wider arc takes the longer time. Yet the times through maximal and minimal arcs of a pendulum up to 5 or 6 feet long differ only by a small fraction of a pulse beat, which Galileo's readers would certainly have called an insensible interval of time, so there was no intentional deception in the *Dialogue*. What Galileo said about pendulums in his *Two New Sciences* will be deferred for discussion in chapter 13.

It long puzzled me why Galileo had *added* times of maximal and minimal swing for some pendulum, getting 1988 grains on f. 189v1, which he then divided by 16, getting 124 *tempi*. His reasoning was probably as follows. As the swings damped down, time to the low point must continually become less. The only reason seemed to be diminishing resistance from the air with shorter swings. Without air resistance a pendulum might never stop, rising ever to the height from which it had first descended. In coming to rest from a maximal swing through air, it would reach some swing from side to side after which there would be no *sensible* difference of times to and from the lowest point, as would be the case in total absence of air.[5] For any pendulum swinging from side to side in air without sensible difference in times of the upswing and the downswing, there would be a longer pendulum which took that same time in the absence of air, through any arc. Galileo appears to have sought such an 'equivalent' or 'ideal' pendulum for the actual pendulum of 1740 *punti* swinging in air.

Pursuing such reflections, I believe, Galileo did the work recorded at the top of f. 189v1, drawing and canceling a diagram and writing and altering a statement. In his units, a pendulum

5   If Galileo had been correct in supposing that air resistance alone was responsible for the difference in time when a pendulum swings through different arcs, he would now have had a way of measuring the effect of air on his pendulums. But, in fact, it is the circular path that is the main reason for unequal times to the vertical for swings of notably different amplitudes. Later Huygens was to utilize this fact in devising his cycloidal checks.

of length 2000 *punti* swings side to side in $2\sqrt{4000} = 126$ *tempi*, almost the same time as his 124 *tempi* for swings through a maximal and then a minimal arc by the pendulum of 1740 *punti*. On the diagram that he canceled a bit later, and in writing the first statement on the page, he entered 2000, making it appear as if that had been the pendulum length, in *punti*, he had timed in the column at top left. But he then gave up 'equivalents' and times from side to side, returning to pendulums timed to the vertical only. Thus it came about that the crucial calculation that he proceeded to make was based only on his actual measurements, with the ratio 942/850 for time of a pendulum swinging to the vertical through a small arc as compared with time of fall through that pendulum length.[6]

That was the ratio which enabled Galileo to discover the law of fall in mean-proportional form, mathematically equivalent to our times-squared law of distances fallen from rest. Had Galileo recognized such a thing as 'the force of gravitation,' and then worked in terms of a local gravitational constant (g, as I call this in his system of units), his *measured* ratio 942/850 would have given $\sqrt{g} = 1.108...$, $g = 1.228...$ In modern dynamic theory, the value of g is independent of any system of units, and would be valid on the moon, or on Jupiter, as well as at any place on earth. That was my basis for stating in the preface that the Galilean units gave g a universal physical meaning. It is the ratio of distance of fall from rest to the length of pendulum timing that fall by swing to the vertical through a vanishingly small arc. So does $\sqrt{g}$ have a universal physical meaning – the ratio of *times* when a fixed length is timed first as a pendulum swinging to the vertical and then as a fall from rest. Or to be exact, that would be true if the measurements by Galileo had been perfect; being measurements, they were not quite exact, for in theory we have: $g = \pi^2/8 = 1.2337...$ and $\sqrt{g} = \pi/2\sqrt{2} = 1.1107.$

---

6 It should be noted that when Galileo checked his timing of 942 grains flow for swing through a small arc, probably of 10° or less, and got 945 grains on f. 189v1, he did not adopt that as a corrected timing. Only the ratio 942/850 exactly agrees with Galileo's crucial calculation on f. 189v1.

The manner in which Galileo arrived at his units excluded his having even thought of determining any *local* gravitational constant of the kind we now use. He did not know enough physics to distinguish Padua from other places on earth in terms of any 'gravitational force,' a dynamic concept. From Euclid, Book v, Galileo knew enough mathematics to discover by painstaking actual measurements that proportionalities indeed exist among distances and times in the physical phenomena of 'natural' motion; pendulums swing and bodies fall with motions beginning spontaneously upon release from restraint. That is one way in which Galileo's mature mathematical physics was more rigorous than various mathematical practices of his time, such as the use of decimal fractions without redefinition of 'number' from Euclid's conception, or the use of algebraic equations that treated $\sqrt{2}$ (for example) as if it were a *number* in spite of Euclid's definition.[7]

Obviously our constant depends on the system of units we adopt, being very different in the British and the metric systems, and capable of being made anything we like (except zero) by a suitable choice of units of distance and of time. Galileo, as nearly as he could measure distances and times in the gravitational phenomena known to him, made it express a proportionality valid anywhere that bodies fall and pendulums oscillate, and in any system of units – including of course his own. It is an amusing fact of history that the noted scientists who were commissioned to establish the metric system came close to doing something that would have had useful practical applications in mental arithmetic when they considered – and rejected – establishing as the meter the length of the seconds-pendulum. (Instead they decided to use one ten-millionth of the length of meridian from equator to pole through Paris.) They knew too much physics to make estimating easy. Galileo knew too little physics to make dynamic distinctions. But he did suggest (in his

---

7 Euclid would have regarded $\sqrt{2}$ as a relation, not as a number; even as a special kind of *ratio*, the mean proportional between 2 and unity. (I say 'unity' because in Euclid's definition, not even 1 was a number; it was the *unit* with which numbers were formed.)

last book) the adoption of a standard unit of velocity based on time of fall through a specified distance (1 *picca*, or about 12 feet).

In both the British and the metric systems the astronomical second is our unit of time. It may seem strange now that Galileo adjusted his original unit of time rather than that of distance, which had been equally arbitrary in the first place. Nothing in mathematics dictated his choice when it came to his altering either the lengths of pendulums or their times of swing in the previous tabulation. But he did own a finely engraved ruler, in units that are about as small as it is convenient to distinguish with the naked eye. That was a very convenient practical unit, especially for Galileo who made all calculations in ratios of integers. Galileo's unit of time had no such practical advantage; no one else ever used it, and I know of no evidence that Galileo himself ever related grains or *tempi* to a second. To calibrate his timing device would have been a great nuisance, to no purpose whatever. Galileo once sent to a friend a length for the seconds-pendulum, for his use in some hydraulic studies, but he discouraged the misguided quest for a *precise* seconds-pendulum by another friend. Better, he said, it was to count oscillations of any long pendulum during the whole time between two successive crossings of the meridian by a particular star, and then to use his mean-proportional rule for timing with that anything in seconds. No length of an actual pendulum would beat exactly one astronomical second, and whatever excess or defect there was must accumulate with beats, as Galileo pointed out. Clocks regulated by pendulums came later, with Huygens.

Returning now to f. 189v1, we have the calculation that put the law of falling bodies into Galileo's hands – or, rather, we have the final step in his series of calculations, of which others had been made on the discarded part of a page when he had adjusted 27,840 to 27,834 on f. 90[b]. The time for which Galileo calculated the distance of vertical fall was 280 *tempi*, double the 140 at which he had timed the 30-foot pendulum that had for him firmly established the generality of the pendulum law. What Galileo calculated on f. 189v1 was one-half the fall in time 280, which he then doubled and entered in a Latin statement: 'The vertical whose length is 48,143 [*punti*] is completed in 280 *tempi*.' In metric units, that says that fall through $45\frac{1}{4}$ meters takes 3.07 seconds; I calculate the time at Padua to be 3.04 seconds.

The final step in Galileo's work looks odd; to get half the fall sought, he multiplied 6700 by 100,000 and divided that by 27,834. If we write this as $6.7 \times 10^8/27,834$, it is easier to see how 6700 represented a ratio rather than a number. Because Galileo did not use decimal fractions, but integers (and ratios thereof), he had to keep track of order of magnitude as he went along, and adjust by a power of 10 at the end. Writing $2T/t$ for 6.7, that was the ratio of double the time for which distance of fall was sought to the time of fall through a known distance, in this instance 4000 *punti*, timed at 1337 grains flow = 83.5625 *tempi* (Galileo wrote 6700 for $6,701\frac{1}{2}$). Because his timing of fall through 4000 *punti* was 1/30 second high, we might expect Galileo's calculation for so long a time as 3 seconds to be very defective. It was not, because in his calculations (using ratios only) t had entered twice. I factor Galileo's procedure as below, after having written, in place of g, the ratio f/p – as Galileo (had he used symbols) would have thought of the general ratio of a distance fallen to the pendulum length swinging to the vertical in the same time. This factoring gives:

$$(Tt/4)(T/t)f/p = 1/2 \text{ fall in time } T = (f/p)(T^2/4);$$

and since $f/p = g$, Galileo's end result was mathematically no different from ours using a local gravitational constant.

Before Galileo noted redundancies in his calculations with ratios, he would never have 'squared a time' (or a measure of time), an operation that as yet made no physical sense.[8] The product Tt above arose by compounding *ratios*, a legitimate opera-

8 Squaring a length did have a legitimate physical sense in terms of areas, and units had long been in use that were analogous to our square foot or square mile. Even now we name no unit the 'square second' or 'square year,' and Galileo did not perform mathematical operations until he perceived them to have a clear meaning.

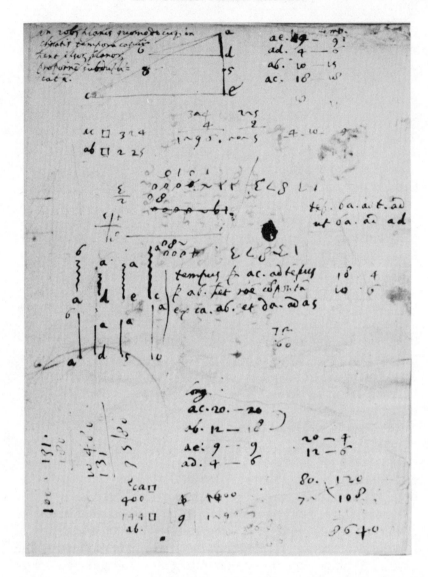

Figure 6. f. 189v2. One entry was made with the page inverted, and before f. 115v (see figure 2, p. 13). This corrected the actual timing of fall through 4000 *punti*. The other entries show how Galileo found the rule for descents along planes that differ both in length and in slope.

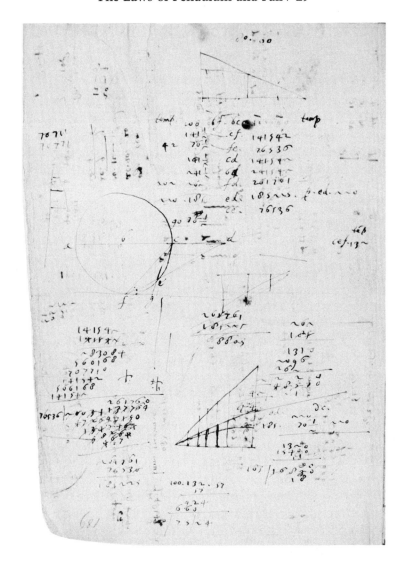

Figure 7. f. 189r. Numerical confirmation of a theorem derived in 1602 on incorrect assumptions. Using the rule found for planes differing in length and slope (figure 6, at left), Galileo vindicated his theorem, for which he later found a valid proof based on uniform acceleration.

tion in Euclidean proportion theory (though Euclid himself used in only geometrically and did not define 'multiply' except for integers). Nor was the ratio *f/p* used directly by Galileo in his (now lost) preliminary calculations; it arose as the product of two occurrences of his ratio 942/850 when calculating from a distance fallen to a length of pendulum, through related times.

Perceiving redundancies in his procedure step by step with ratios, Galileo next took directly the mean proportional between the distances 4000 and 48,143, recognized this to be almost the half of 27,834, and drew at lower left on f. 189v1 the diagram that he used from that day on to relate times and distances in fall by the mean-proportional rule. On f. 189v2 he used his new found law to correct his one poor timing (for fall through 4000 *punti*) from $83\frac{1}{2}$ *tempi* to $80\frac{2}{3}$, almost exact. Then on f. 115v he summarized what he had found, this time writing 48,143 p. (for *punti*), confirming my initial assumption that the same unit had been used in 1604 that was first identified from later notes of Galileo's.

Next, having found the law of free fall, Galileo wondered whether it held also for descent along an inclined plane. Taking up again f. 107v (from which we started), he entered in the margin of his tabulation (distances rolled after each of eight successive equal times) the first eight square numbers. Each, multiplied by the first distance, gave the corresponding measured roll almost exactly. Returning to f. 189v2, Galileo conjectured a compound ratio for descents along planes differing both in slope and in length, found it defective, and by arithmetical test found the correct rule. Using that rule, he calculated on f. 189v his arithmetical confirmation of a theorem he had sent to Guidobaldo del Monte in 1602, further discussed near the end of chapter 4.

It is no wonder that Galileo preserved all his life f. 189v1, on which he had arrived at the times-squared law of fall. That led him at once to a host of further theorems and solutions of problems, to be traced later on in this book. Without Galileo's discarded papers, f. 189 was a difficult document to understand fully, but it then became the key to Galileo's mature mathematical physics.

What had preceded f. 189 (and f. 107) in his gradual path to modern science will be considered after a historical interlude shows the anachronism of Galileo's law of fall by providing information about its antecedents and its immediate reception among seventeenth-century natural philosophers.

CHAPTER TWO

The Medieval Context

Galileo's law of fall was an anachronism in 1604. As is evident from the preceding chapter, its historical background need not be known in order to understand the work that put the law into Galileo's hands. Hence the present chapter is in the nature of an interlude that may be skipped over without effect on the biographical contents of the main text.[1]

The medieval concept of impetus, by which fourteenth-century natural philosophers had arrived at a different mathematical rule for distances and speeds in fall, was certainly known to Galileo when he began his study of natural motions. Impetus theory was revived and put forth as *underlying* the times-squared relationship among measurements made by Galileo, after he died in 1642. Scientists may find this story interesting because it reveals an unexpected early concept of quantum-jumps in competition with the rival idea of mathematically continuous change.

Summarizing the history of fall theories from the fourteenth to the seventeenth century, a distinguished scholar wrote thirty years ago:

In the first moment after the removal of a body's support the impetus was held to result from an external cause – gravitational attraction. During later instants, however, the impetus itself was taken to cause movement. Temporal and spatial aspects of impetus were clearly distin-

1 Most of this chapter was anticipated in my 'Free fall from Albert of Saxony to Honoré Fabri,' *Studies in History and Philosophy of Science* 5 (1975), 347–66.

guished by da Vinci: 'The freely falling body acquires with each degree of time a degree of motion, and with each degree of motion a degree of velocity.'

Why did Leonardo, Benedetti,[2] and Varro assert the proportionality of the falling body's velocities to the spaces traversed, and not to the times? Doubtless they regarded these as equivalent ... Given the alternatives, (a) velocities are proportional to the times, and (b) velocities are proportional to the spaces traversed [from rest], Leonardo, Varro, Galileo (1604), and Descartes all chose (b).[3]

Until 1974 I considered Hanson's question legitimate and his answer appeared to be the only plausible one. His question was, however, based on a misapprehension that is still shared by most scholars.

It is true that Michael Varro, writing while Galileo was still a student,[4] stated that speeds in fall are proportional to the distances from rest, in the ordinary sense of 'proportional.' It may be that others before him had said this explicitly, but not Leonardo or anyone else I have read, nor is it necessarily implied by anything said before Varro. The accepted form of a mathematical rule for speeds in fall before Galileo is traceable to Albert of Saxony and was rooted in a totally different conception of proportionality from ours.

In order to make this vital point quite clear, let me begin by substituting *unit* for *degree* in Leonardo's statement. His own word *grado* originally meant a step, as on a ladder or stairs; that is, a distinct and countable thing in a discrete series. Speeds were traditionally numbered by integral degrees. Let us read Leonardo as saying 'The freely falling body acquires with each unit of time a unit of motion, and with each unit of motion a unit of velocity.' One kind of 'proportionality' among times, distances, and speeds is implied when these units are taken in succession, but not the proportionality existing when they are taken cumulatively from rest. The data for a falling body might be tabulated thus:

---

2 See chapter 3.
3 N.R. Hanson, *Patterns of Discovery* (Cambridge 1958), 39
4 Michael Varro, *Tractatus de motu* (Geneva 1584), 14

During time 1, it falls a distance of 1 with a speed of 1

2   2   2

3   3   3

and so on. The acceleration would be uniform but not continuous. Equal speeds are added in equal times, but discretely at the beginning of each new period of time, a speed remaining uniform during each period. A proportionality might be said to exist between such speeds and the corresponding times from rest, but not a proportionality in the algebraic sense when time and speed are both considered as undergoing continuous change. In any case a proportionality could not exist between speeds and distances in the same way, for although the numbers representing times and distances are the same, they do not accumulate in the same way as do times and speeds. In medieval terms, times are successive but distances are 'permanent' – that is, capable of being present all together, unlike instants. Thus by the end of time 2, the distance has become 3, and by the end of time 3 the distance has become 6. In short, no proportionality (defined by Euclid as 'sameness of ratio') could exist between speeds and distances in fall, as it could exist between speeds and times.

There are good reasons, physical as well as mathematical, for believing that writers before Varro thought of free fall in the way represented above. Hence it is dubious that any medieval writer thought of speeds in fall as proportional to distances from rest. Very probably they regarded speeds in fall as discrete, uniform, and successive, and thought of both time and distance as continuous. In the course of this chapter it will be seen how, *after* Galileo's analysis of speeds as changing continuously, the older concept found acute and eloquent defenders, lending support to the foregoing analysis.

Euclid treated of proportion among continuous magnitudes in Book V, and among arithmetical quantities in Book VII. An omitted definition in Book V and a spurious medieval definition there led to emphasis on *numerical* relations in medieval proportion theory, which was incapable of representing the continuum adequately. Physical entities such as times and distances could be taken in units as small as one pleased, but remained durations and extensions not reducible to mathematical points without loss of physical significance to natural philosophers.

This mathematical approach was in exact accord with certain physical conceptions of Aristotle. It was impossible in Aristotelian physics for a mathematical instant to be both the end of rest and the beginning of motion, or there would be an instant at which the same body was at rest and in motion, against Aristotelian logic. A 'physical instant' could exist, having a duration however small, and though that phrase came into use only later, it throws light on medieval physics conceptually, as in the doctrine called 'latitude of forms' from which the so-called mean-speed theorem evolved.[5]

Albert of Saxony mathematicized fall on the basis of the medieval impetus theory developed by Jean Buridan in the fourteenth century. Increase of speed, he reasoned, might be by addition or by multiplication (an operation defined only for numbers). The possibility of infinite speed gained in a finite time was denied, and Albert concluded:

It is understood that the speed is increased by double, triple, etc. in such a way that when some space has been traversed by this, it has a certain velocity, and when a double space has been traversed by it, it is twice as fast, and when a triple space has been traversed, it is three times as fast, and so on.[6]

Historians assumed that distances and speeds in fall must be measured from the beginning of motion, and that Albert intended a continuing uniform acceleration which would make speeds in fall proportional to distances traversed. But it is more probable that Albert alluded to *successive* spaces and velocities, in which case his rule was that the falling body goes a certain distance at unit speed, and *then* twice as far at double speed, and *then* three times as far at triple speed. Since n-fold distance at n-fold speed

5 Properly a middle-degree postulate, this was never applied to an analysis of motion in free fall because impetus theory excluded fall as a case of uniformly accelerated motion; see below.
6 See Marshall Clagett, *The Science of Mechanics in the Middle Ages* (Madison 1959), 566, 568.

implies equal time increments, the foregoing table probably represents Albert's rule correctly.

That is a rule very simple to grasp, and it is likely that others, including Leonardo, accepted it as the mathematics of speed in fall. Varro, late in the sixteenth century, instead adopted literal proportionality of speeds to distances from rest – an untenable assumption, as Galileo showed in *Two New Sciences*.[7] No doubt medieval mathematicians would have quickly exposed its untenability if Albert, or anyone else, had proposed it earlier. Under Varro's rule, a falling body could not start from rest. Albert's rule, unlike Varro's, may in fact hold for actual fall; at any rate, quantum-jumps of speed cannot be excluded as a logical possibility.

This possible applicability of the impetus-theory rule to actual fall was sensed not long after Galileo's death by his friend Baliani, who had published a book including the times-squared law in 1638, the same year that *Two New Sciences* appeared. Baliani well knew that measured distances in successive equal times increase as do the odd numbers, but in his second edition eight years later he suggested that behind this rule there may lie another – the rule I ascribe to Albert of Saxony, applied to insensible distances growing as do the natural numbers.[8] His reasoning can be paraphrased as follows:

Take three distances from rest in successive equal times of fall. Suppose each to be divided into ten parts, not equal but progressing as the natural numbers, and assume these smaller distances to be passed in equal times. The first ten such distances are traversed in the same time as the second ten, and the third, while the sums of the distances will be 55, 155, 255, or nearly as 1:3:5. Using a hundred divisions instead of ten, the numbers would be 5050, 15050, 25050; and the relation 1:3:5 would be borne out to any desired approximation for distances covered in 'physical instants' millions of times briefer than we can measure.

---

7   There he showed that it requires an initial discontinuity between rest and motion.
8   G.B. Baliani, *De motu naturali gravium solidorum et liquidorum* (Genoa 1646), 111–13

In principle, a quantum-jump account of acceleration in fall, with successive extremely brief uniform speeds each greater than the preceding, and contiguous so to speak, is just as viable as the continuous growth of speed adopted by Galileo and by us. The calculus makes Galileo's the simpler assumption to work with. In Albert's time no one calculated distances, times, and speeds in fall, but thinking abstractly about them was facilitated by the medieval arithmetical theory of proportion ingeniously developed for lack of the rigorous Euclidean theory in Book v.

As I interpret Albert's rule, it does not involve any internal contradiction, as did Varro's idea (proportionality of speeds to distances fallen from rest, in which a body at rest could never start falling without a discontinuity). All that was wrong with Albert's rule was that it does not conform with actual measurements. It implied that in the second of two equal times, beginning from rest, a body falls twice as far as in the first of those times, whereas measurement shows it to fall three times as far. That faults Albert's rule as nature's law, but it does not fault Albert as a natural philosopher at a time when recourse to actual experience was no part of physics.

Had anyone measured distances in natural descent from rest during a time and its half, finding the ratio to be 3:1, fall would have been identified at once as a case of 'uniformly difform motion' as uniform acceleration was then called. That ratio was known to be a characteristic of such motion, by the mean-speed postulate previously mentioned. Without measurements, it sufficed to exclude fall as a case of uniform acceleration, since the ratio in fall had been put at 2:1. Hanson's question accordingly answers itself, for in uniform acceleration the speeds acquired are indeed proportional to times from rest. The same could not be true in fall under impetus theory, so no one even raised the question of time-proportionality that has seemed so natural to modern historians.

Galileo first announced his times-squared law of distances in fall from rest in his famed *Dialogue* of 1632, where he did not supply a rigorous derivation of it, but only what he called a 'probable argument' for its corollary – that in uniform motion following fall, another equal time will carry the body double the distance already traversed. Descartes read the *Dialogue* and wrote to his friend Marin Mersenne:

But I must declare to you that I have found in his book some of my thoughts; among others, two that I think I have written to you before. The first is that the spaces through which heavy bodies pass when they fall are to one another as the squares of the times they take in descending; that is to say that if a ball takes three moments to descend from *A* to *B*, it will take only one [moment] to continue on to *C* [twice as far], etc. This is what I said [but only] with many restrictions, since in fact it is never entirely true, as I am thinking of proving. (Hanson, 49; see note 3)

Apart from the fact that Descartes did not believe Galileo's law or its corollary of double-distance to be ever exactly true, it is of interest that he cited only one special case, in this letter and elsewhere – the same unique case that was invariably discussed during the Middle Ages. That was the bisection of a completed fall, rather than its division into any number of equal parts, or the indefinite continuation of a fall as dealt with by Galileo in *Two New Sciences*. The most that could be implied by the case of bisection was the result of repeated bisections in the direction of the beginning of fall, leaving an infinitude of other cases completely uninvestigated. All cases are included only under the assumption that change of speed is mathematically continuous, an assumption not made by natural philosophers who were seeking the *cause* of acceleration in fall. An infinitude of causes during the finite motion was not permissible. Moreover, Aristotle had banned the category 'change of change,' which would have included acceleration, motion being change of place with respect to time.

Now, seeking the cause of natural phenomena was the only purpose of science as Aristotle defined it, and philosophers did not abandon the quest for the cause of acceleration in fall just because Galileo ridiculed it in *Two New Sciences*. Descartes was not alone in rejecting Galileo's law of fall. Pierre Cazré rejected it on one basis; Honoré Fabri on another. Pierre Fermat offered a mathematical proof against it before 1638, and afterward gave a demonstration against the possibility that speeds in fall are proportional to distances from rest.[9]

Essentially new in Galileo's theory of fall were the passage from a quantum-concept of speeds to the concept of continuous change, and the abandonment of the causal quest as superfluous in his new science of motion. Re-examined in the above light, the preceding medieval work was more complete and consistent than it has previously appeared. It continued to influence physicists in the seventeenth century, with the exception of Galileo from 1604 on, though it is precisely Galileo whom historians have portrayed as indebted to it for his mature physics.

The most illuminating book with regard to medieval tradition after Galileo's death is Honoré Fabri's *Tractatus physicus de motu locali* ... , published at Lyons in 1646. Fabri's axioms were that impetus is additive; that the ratio of increase in cause is that existing in the effect; that the same cause produces equal effects in equal times; and that a cause acts in inverse ratio to resistance. Many of his theorems agreed with Galileo's while many others contradicted his. Whereas Galileo postulated that equal times in fall produce equal speeds, Fabri offered a proof: motion increases proportionally to impetus, while equal degrees of impetus are acquired in equal times.

Fabri's physical concepts reflect unspoken assumptions made by many of his contemporaries and probably by all his medieval precursors. To his theorem that impetus increases by single instants which are all equal, Fabri added in a scholium:

Note above *equal instants*, since physical time cannot be explained except by finite instants, as shown in my *Metaphysics* [Lyons 1647]. But however this may be, I call an *instant* all that time, whether large or small, in which something is produced all at once. Whether a larger or a smaller part [of time] does not matter for our purposes, since given any finite time you can have a larger and a smaller [time]; that is certain. Therefore all that time in which the first impetus acquired is produced

---

9 Galileo considered a single sentence (or two) sufficient for any attentive reader of his *Two New Sciences*. Le Tenneur perceived that it was; Fermat's demonstration took half a dozen pages.

is called 'the first instant of motion,' to which there then succeed equal times thereafter. (*Tractatus physicus*, 87)

Thus, a first *physical instant*, as Fabri later calls these, exists, in which a first degree of impetus is produced; there is no continuous growth of impetus from rest. Here Fabri's physics departs from Galileo's, returning to the medieval physics from which Galileo had departed. Once the notion of a 'first instant' was re-established, the rest was easy for Fabri. A scholium to Theorem 44 explained the sense in which proportionality can be said to exist between distances and speeds in fall, not at all the sense required for continuity, but strictly numerical:

From this I may say that speeds grow in every instant according to the ratio of spaces run in the instants.

This is certainly true when the legitimate senses of the words are understood, though Salviati[10] cries out in Galileo's Third Day when the progress of growth [in time] by individual instants is assumed – 'Why at one [instant] rather than another?' (ibid., 88)

Salviati was of course alluding not to 'physical instants' but to mathematical instants, dimensionless points in time:

Certainly if the speed in one instant be compared with that in another, the two will be in the same ratio as that of the spaces [!]; if therefore in one instant one space is run through with one degree of speed, surely in [two] equal instants there is acquired double space at two degrees of speed.' (ibid.)

So that we have in Fabri exactly the rule previously tabulated for Albert of Saxony:

... nor does it hurt, as Galileo objects, that then these motions are uniform, for motion made in an instant *should* be considered as uniform ... Nor is it any bar when he says that instantaneous motion

10 Galileo's spokesman in *Two New Sciences*, criticized by Fabri.

cannot exist ... I explained in my *Metaphysics* the extent to which instantaneous motion can exist. (ibid.)

Unlike Baliani, who published his rationalization of the medieval rule as lying behind the times-squared law in the same year as Fabri's book, Fabri denied the odd-number rule and rejected Galileo's statement that to reach any speed from rest, a body must pass through every possible lesser speed.

Galileo is refuted here on two grounds. First, because he unnecessarily assumes infinitely many instants, and second, because the ratio he gives is unconvincing. He calls rest 'infinite slowness,' and no doubt *his* motion would be propagated through every degree of slowness. But against this, first, rest is not in fact slowness, since rest can have no motion. Second, fast motion immediately follows after rest as well as does slow motion, as is evident from projectiles. Third, motion does begin, and hence with something, so the first motion is not infinitely remote from rest [as Galileo asserts].

I reject the opinion of those who want the natural acceleration of motion to be such that the spaces acquired in equal [successive] times follow the odd numbers 1, 3, 5, 7, etc., with the spaces [from rest] as the squares of the times ... That is obviously false from what has been said, and certain if in equal times equal moments of speed are acquired. (ibid. 96–7)

The last sentence shows the abyss between Fabri and Galileo, for it was precisely on that definition of uniform acceleration that Galileo established the odd-number rule. Fabri's arguments show the vitality and unbroken persistence of medieval impetus theory. By the same token, they show the complete independence of Galileo's mature physics from medieval influence, despite his acceptance of impetus as a kind of impressed force in his early writings on motion, to which I shall turn after one concluding comment on Fabri's natural philosophy.

Baliani was himself an able experimentalist. Fabri appears not to have made actual measurements, but he did not challenge Galileo on that score or question the accuracy of his statement that in a hundred trials no variance of more than one-tenth of a pulse beat from the times-squared law had occurred. What was at issue was the conceptual basis of science – how the natural philosopher should imagine events lying beyond the reach of direct measurement. The issues for Fabri are stated in the following extracts from his long scholium on naturally accelerated motion.

The opinion thought out by Galileo is confirmed by his supporters on two heads; first, by experience, and second, by reason. As to experience, three very powerful experiments are offered. The first is by motion vertically downward ... for indeed many grave writers on philosophical matters have often tested this sensibly, repeating their experiments *ad nauseam* ... The second experiment is on the inclined plane ... and Galileo says expressly that he himself tested this often and found no discrepancy ... The third experiment is taken from pendulums, in which it is observed that the lengths are as the squares of the times ... On these three very powerful experiments is founded Galileo's hypothesis, which in my opinion could not be more clearly or more sincerely expounded [than I have done].

Before adducing the reasons behind this opinion and refuting them, I shall first show how our hypothesis can stand with those experiments. Surely there is a rule about this that no philosopher denies; namely that whenever some experiment is such that two contrary hypotheses can stand with it, the experiment can be deduced to be neutral; therefore I shall show that Galileo cannot legitimately deduce his hypothesis from the said experiments.

Taking either some instant in time or some distance completed or run through, if one says that in fact the second distance is triple the first minus one in a million points, or that the second time is greater than the first by one in a hundred thousand instants, how is the difference in either space or time to be sensibly perceived? For every physical experiment must be subjected to the senses ... I ask Galileo whether he, or anyone else, can say whether one space is triple another, and whether if someone says that it is off by one-millionth, the experiment carries conviction?

The first reason Galileo gives in favor of his law is that since Nature always uses the simplest means in her operations, and since acceleration in natural motion cannot be made more easily and simply than in this progression of squares, there is no doubt that natural acceleration is made in that ratio, especially when all experiments agree with it and it can explain all the phenomena.

I reply that the first arithmetical progression, in the numbers 1, 2, 3, 4, is much simpler than that by 1, 3, 5, 7, as no one would deny. Next, since two hypotheses happen to agree with all the experiments or phenomena, there must be some reason why one should be chosen over the other; but there is no reason why Galileo takes his, as we shall see. But we prove our ratio demonstratively; therefore ours is to be preferred as the true theory of things. Yet because the other approximates for sensible times to the true one, it may be taken in practice and for the ordinary measures of these motions.

Galileo's hypothesis is false under the hypothesis of finite instants, since new increments of speed are made in single instants; but physically speaking it comes out the same as if it were true, because it can be tested only in sensible parts of time. Surely, since any sensible part contains almost innumerable instants in which the progression takes place, the difference between the two [hypotheses] cannot be made sensible ... Thus, in the common opinion in which it is said that time consists of actual infinite parts, Galileo's progression can stand. Therefore, here is the key to the difficulty: the simple progression has its principle physically, not experimentally; the odd-number progression, by experiment and not principle. We combine the two, by principle, and by experiment. (ibid., 99–108, passim)

Elsewhere Fabri remarked that Galileo's theory and his own agree in observed results, but the former cannot give the cause of acceleration while the latter can and does, so it is superior. In fact, by ancient and medieval standards, Galileo's physics was not a true science; intelligent traditionalists like Fabri used the same standards, and so do many philosophers today. Descartes said that Galileo had built without foundation because he did not start from the cause of motion. A similar objection was made against Newton's law of universal gravitation, for which Newton himself felt obliged to seek a cause (though in vain). Like the law of fall, Galileo's entire outlook in physics was historically an anachronism; it is not to be found in the history of earlier natural philosophy.

CHAPTER THREE

Galileo's Pre-Paduan Doctrine of Motion

Galileo entered the University of Pisa as a student in 1581, remaining there four years. At first he studied medicine, but in 1583 he became more interested in mathematics and natural philosophy (as physics was traditionally called). He left without a degree, and during the years 1585 to 1589 he tutored privately at Florence and at Siena. At Siena, in the academic year 1586–7, he also held a public teaching post. From 1589 to 1592 Galileo was professor of mathematics at his alma mater, with the obligation of teaching Ptolemaic astronomy and Euclid's *Elements*.

In 1584, while still a student, Galileo composed a set of lectures *De universo* in the hope of teaching natural philosophy, his father having told him that he would not support another year of university study. The lectures are Aristotelian in content.[1] Galileo explicitly rejected the Copernican system in his early notebooks, on the grounds that the earth must be immovable at the exact center of the universe, as proved by philosophers and astronomers.

Galileo's first scientific composition, of 1586, was in Italian and was called *La bilancetta*, literally 'the little balance,' though what it concerned was by no means small. It was a hydrostatic balance having arms each two feet long, and very sensitive indeed

---

1   An English translation of *De universo* was published in William A. Wallace, *Galileo's Early Notebooks* (Notre Dame 1977), 25–160. On pp. 161–251 is Father Wallace's translation of the sequel in 1588 (on qualitative change and on the 'elements'), mentioned later in the present chapter. Wallace dates both manuscripts to Galileo's years as professor at Pisa (after 1589).

(the diminutive in its name applied not to the instrument, but to the weights it was capable of detecting). The treatise reconstructed the procedure of Archimedes in solving the famed 'crown problem' of analyzing a gold-silver alloy without damaging the object to be tested. This composition had been preceded in 1585 by specific-gravity determinations for various metals and gems that show great skill and care on Galileo's part. His weighings in air and water were recorded in fractions of one grain, and the results were in good agreement with modern tables of specific gravity. Variance would be expected with errors of even one or two grains in the weights of Galileo's gemstones.

To insure precision, Galileo wound fine wire tightly along the arm from which the counterweight was hung. He counted the turns from center to suspension with the point of a stiletto, using his ear as well as his sense of touch, as he wrote in his treatise. It is accordingly safe to date Galileo's first measurements to the latter part of 1585, not long after he had left the university. His interest in Archimedes had been sparked by his teacher of mathematics, Ostilio Ricci, said to have been a student of Niccolò Tartaglia at Venice. Neither Tartaglia nor Ricci ever taught in a university; Ricci was court mathematician to the Grand Duke of Tuscany, a post later held by Galileo at Florence. There is no doubt that Galileo was taught Euclid's *Elements* from Tartaglia's Italian translation, the first version in Europe to comment correctly on the general theory of ratios and proportionality among mathematically continuous magnitudes, in which theory Galileo's mature physics was firmly based.

Tartaglia once promised an accurate solution of the crown problem, but never published it. Galileo's solution was simple, straightforward, and exact, unlike the two accounts of Archimedes' procedure[2] that had been generally accepted for centuries. Galileo based his approach on authentic propositions from the work of Archimedes *On Bodies in Water*. Tartaglia had first

---

2   One of them depended on measuring overflow of water from a full container upon immersion of the alloy and of specimens of the component metals. The other required weighing the alloy and an identical replica of it in pure metal, not very practical for a king's crown made by an artisan. Archimedes did not publish his solution, which Galileo reconstructed in *La bilancetta*.

printed its medieval Latin translation in 1543, and then his own Italian version, with commentaries, in 1551. This book, which was reprinted several times, became of particular importance to Galileo's early work, for in it Tartaglia had made a statement – on the very first page (and again in the commentaries mentioned) – that he himself did not follow up. He had said that speeds of bodies sinking through water are as their specific gravities.

Tartaglia's statement turned Galileo's interest from hydrostatics to criticisms of Aristotle in the matter of natural motions. A former student of Tartaglia's had done similarly in 1553, publishing the first demonstration that speeds in fall are not dependent upon weights, as Aristotle had taught. This was G.B. Benedetti, who in the following year published a book on motion 'against Aristotle and all the philosophers' that was plagiarized in Belgium and came to the attention of the Dutch engineer Simon Stevin. In 1586, just as Galileo began his own scientific career, Stevin published the results of tests from a height of 30 feet, finding that bodies very different in weight fell with equal speed. Meanwhile, in 1585, Benedetti had also reprinted his new theory of natural motions along with numerous other works of his on mathematics, mechanics, and science.[3]

This new stirring in Europe outside the universities, which had long had a virtual monopoly on physical science as a branch of philosophy, is traceable mainly to the work of Tartaglia (and, through him, back to Archimedes). Tartaglia was a self-taught mathematician who contributed to algebra the solution of cubic equations and restored the classical rigorous theory of ratio and proportion, lost to Europe for centuries by reason of an Arabic corruption of Euclid, Book v. The new movement in physics was ushered in by a mathematical theory depending upon concepts neglected in the existing university tradition.

It was Stevin, in 1585, who introduced the use of decimal

---

3 Galileo's demonstration from the Leaning Tower of Pisa did not depend on Stevin's book, printed in Dutch, and it is unlikely that he had seen Benedetti's 1585 book. His own 'Pisan *De motu*' will be discussed presently; it contained his reasons for speeds of fall unaffected by weights.

fractions; his book on arithmetic defied even Euclid by opening with a chapter titled 'One is a Number.' Today that title seems trivial and obvious, but in 1585 its implications were so revolutionary that a talented French mathematician – the only one outside Italy who truly understood Galileo's mature doctrine of natural motion – later declared that Stevin's dictum marked the end of true rigor in mathematics.[4] In a sense it did, until the redefinition of 'number' by Richard Dedekind. But it also opened up a whole new era in practical mathematics and in physical theory, resulting in a period of unprecedentedly rapid discovery in science.

I mention these things not because Galileo took up the new mathematics, for he did not. He never used algebra, or even decimal fractions. My reason for describing the situation when he began his scientific career is to make it clear that much was in progress after Tartaglia to which historians of physics have paid scant attention, perhaps believing that philosophy must have been the sole source of the Scientific Revolution. Nothing could be further from sober facts, as one might easily guess from the attacks leveled by academic philosophers of his time (and long afterward) against everything Galileo wrote. Galileo was not ignorant of the new developments in mathematics that he did not apply in his own physics. He did not apply them because he was aware that they lacked rigor, as was the French critic of Stevin and supporter of Galileo, Le Tenneur.

While Galileo was teaching at Siena in 1586–7 he composed a Latin dialogue, untitled and left unfinished when he returned to Florence in mid-1587, which might suitably be called 'Problems of Local Motion.' It was a teacher-pupil dialogue, the form that Tartaglia had used for his 1551 commentaries on the hydrostatics of Archimedes, as well as for his 1546 commentaries on medieval statics. Both those works had been reprinted together with the 1537 *New Science* by Tartaglia, after his death in 1557,

---

4   This was Jacques-Alexandre Le Tenneur, who in 1640 published anonymously at Paris a book criticizing the innovation by Stevin. In 1649 he published a vigorous defense of Galileo's analysis of uniform acceleration in fall against the attack by Honoré Fabri, who revived medieval impetus theory in 1646; see chapter 2.

and it is hard to doubt that Galileo read them. This 'new science' was Tartaglia's pioneer application of mathematics to artillery practice, a topic later taken up by Galileo. In his 1546 book Tartaglia described the *squadra*, an instrument for measuring the heights of distant targets; Galileo later combined that with Tartaglia's elevation gauge for cannons when devising his own 'geometric and military compass' (and thus creating the first general-purpose calculating instrument). The degree to which Tartaglia's writings inspired Galileo and others during the early stages of recognizably modern science is missed by those who suppose science always to have depended only on philosophy.

In Galileo's dialogue on motion he mentioned the *bilancetta* he had devised and supplied in full his solution of the Archimedean crown problem, saying that in arriving at this solution he had demonstrated a number of Archimedean theorems 'more physically and less mathematically' than Archimedes, using assumptions more evident to the senses. He then set forth his own new demonstrations as the basis of his non-Aristotelian approach to the subject of motion. That was non-Aristotelian because it was a mathematical approach, contrary to Aristotle's quite logical declaration that the method of the mathematicians could not be the proper method of physics as the science of nature, because mathematicians abstract entirely from matter, while everything in nature possesses matter. How Galileo responded to that, right from the outset, can be seen from his above statement and procedure. By using assumptions evident to the senses, he introduced physical phenomena, and those would not cease to be implied in valid deductions, whether logical or mathematical. He agreed with Aristotle that there must be more to physics than pure mathematics, but he did not agree that it could be in any way improper to employ in physics the *method* of mathematicians.

A single postulate opened Archimedean hydrostatics – that a solid placed in water would move if it were not pressed equally in every direction. That is indisputably true by the principle of sufficient reason, so in effect Archimedes made a concealed definition rather than a physical assumption. It could not be verified by experiment, nor did it have to be, because to deny it would be foolish; it was, in a way, built into language itself. Galileo replaced it by examining the behavior in water of solids whose weights are equal to the weight of equal volumes of water, and then of those whose weights exceed, or fall short of, that of water. In other words, Galileo brought in physical phenomena of motions, and used those to arrive at the same theorems Archimedes had demonstrated by conditions of equilibrium alone. Rising or sinking in water is sensibly seen, in a way that pressures cannot be, rendering claims about rising, sinking, and floating subject to direct experimental test – and requiring such a test to possess physical truth in Galileo's non-Archimedean and non-Aristotelian approach.

In the course of his dialogue on problems of motion Galileo arrived at a curious view of free fall that for many years stood in his way in the crucial matter of acceleration. It arose from the custom of seeking causes, the sole purpose of physics in the Aristotelian natural philosophy he had learned as a student. For Aristotle, science was the understanding of things in terms of their *causes*. A very good causal explanation of acceleration in fall had been formulated in medieval impetus theory,[5] and G.B. Benedetti had followed that lead. Tartaglia had not. Neither did Galileo, who supposed that when a heavy body is thrown up, the cause of its upward motion is an 'impressed force' that diminishes as the body rises. When it reaches its highest point, the 'impressed force' is not exhausted, but is no longer strong enough to move the body upward, so that as it begins to fall it is impeded until all the contrary 'impressed force' is used up. It then moves at the unique speed appropriate to the material of the body in air, as implied by Tartaglia's statement about speeds of sinking in water. A heavy body supported at a height has that same amount of 'impressed force' upward, imparted by the support, as Galileo added in 1590.

5   In the fourteenth century Jean Buridan formulated this theory and gave an explanation of acceleration in fall that Albert of Saxony then interpreted mathematically. In the 'first instant of motion' the body was moved only by its natural tendency downward, acquiring simultaneously one degree of impetus. In each ensuing instant, the natural tendency acted together with accumulating degrees of impetus. This was, so to speak, a quantum theory of speeds in which each was uniform and faster than the preceding. It followed that in fall, acceleration was uniform,

In that view, acceleration is a temporary condition at the beginning of fall from rest, not a phenomenon that continues to persist indefinitely. Galileo's unfinished dialogue ended with a reflection about the optics of observing long falls from a high tower and the possibility that belief in continuing acceleration is rooted in optical illusion. On his return to Florence he read of a similar explanation of acceleration in fall attributed to Hipparchus in antiquity. Instead of completing his dialogue, he began writing memoranda on motion, among the first of which there is a long passage in dialogue form, marked for insertion at the appropriate place. Galileo continued adding memoranda for about a year and then composed a new treatment, in lecture form, that may be called 'Natural Order and Local Motion.'

Meanwhile Galileo's interest shifted to centers of gravity of solids, concerning which Archimedes doubtless wrote, though only his theorems on plane equilibrium survive. Federico Commandino revived the study about the time Galileo was born. In the latter part of 1587 Galileo devised an ingenious approach to centers of gravity in parabolic conoids that he hoped would win him the vacant chair of mathematics at the University of Bologna. He took his basic theorem to Rome, leaving it with the famed Jesuit mathematician Christopher Clavius, whose favorable opinion would weigh heavily with university authorities. He also sent a copy to Guidobaldo del Monte, author of the most advanced Renaissance treatise on mechanics, and he showed another copy to Giuseppe Moletti, professor of mathematics at Padua, whom Galileo was destined to succeed a few years later.[6]

Moletti recommended Galileo for the chair at Bologna, but it was awarded to the Paduan astronomer G.A. Magini. Both Clavius and Guidobaldo had had doubts about the validity of Galileo's theorem, so he sent a supporting lemma. Clavius thought that it begged the question, but Guidobaldo had already resolved his own doubts even before he saw the new lemma. The geographer

but not continuous. The concept of impetus was that of an 'impressed force,' as will be seen next.

6 Concerning Moletti, see W.R. Laird, 'Giuseppe Moletti's *Dialogue on Mechanics* (1576),' *Renaissance Quarterly* 11:2 (1987), 209–23.

Abraham Ortelius was shown the theorem at Rome and sent a copy to Michael Coignet at Antwerp, who at once wrote directly to Galileo to compliment him not only on his work, but also on the improved practical utility of his approach in contrast with that of Commandino. This was Galileo's first recognition from abroad, and any reader who consults his treatise on centers of gravity will recognize its originality.[7]

In 1588 Galileo again took up the project he had started as a student, composing lectures to follow *De universo* in the usual order of the course on natural philosophy as given at the Jesuit Collegio Romano, where he had visited Clavius. It is probable that he there met Paul Valla, hearing some of his lectures on logic, and carried back to Florence a copy of Valla's previous lectures on natural philosophy. Those guided his writings on qualitative change and on the 'elements' (earth, water, air, and fire). The next topic in order would be local motions of the heavy and the light.

Galileo, dissatisfied with the customary Aristotelian treatment, proceeded to write his own lectures on that subject. He opened these with a section on the natural order of the elements, in which motion was not so much as mentioned. Then followed Galileo's 'hydrostatic' account of motion, as in the unfinished dialogue, but differently treated, with emphasis on philosophical rather than practical problems. The lectures on motion followed a theme found about midway in Galileo's old memoranda that had been written over a period of about a year.

At the end of 1588 Galileo received, probably from Valla at Rome, the full lecture notes of the course in logic just ended, which contained much on the logic of demonstration in science. Galileo was so much interested in this that he made a copy for himself, editing and modifying Valla's text as he went along, early in 1589. Later that year he became mathematician at the University of Pisa, and for the first time he studied at first hand both the *Almagest* of Ptolemy and *De revolutionibus orbium*

---

7 It was published as an appendix to Galileo's last book; my English translation will be found in Galileo, *Two New Sciences* (Toronto 1989); see also S. Drake, *Galileo at Work* (Chicago 1978), 13–14.

coelestium by Copernicus. Galileo's knowledge of astronomy had previously been limited to the standard university textbook by Clavius, from which he had quoted in *De universo*, but more than that was expected of a public professor. Galileo was won over to the planetary astronomy of Copernicus, with the sun instead of the earth at the center of the celestial motions, but did not yet accept the earth itself as a planet. Such a curious and hybrid geo-heliocentric astronomy had already been proposed in 1588 by Tycho Brahe in Denmark, but Galileo did not know that.

The philosophical idea that the earth must be at the exact center of all celestial motions was so strongly held among Greek intellectuals that when, about two centuries after Aristotle, accumulated astronomical measurements had shown it to be false, Greek astronomers were warned to leave natural philosophy to the philosophers. Ptolemaic astronomy, with its eccentric deferents, epicycles, and equants, was regarded by philosophers as a fiction made up of purely mathematical hypotheses adopted to 'save the phenomena,' not as a serious description of actual motions of physical entities. When Copernicus implied real physical motions of the earth, he defied that very ancient and rigid tradition, though a spurious foreword printed with *De revolutionibus* in 1543 pretended the contrary. Tycho became famous by outlining the hybrid astronomy that Copernicus might perhaps have settled for, had he subscribed to the tradition that astronomy must have nothing to do with physics.[8] Doubtless Galileo expected to gain fame as an astronomer by presenting the hybrid scheme, unaware that anyone else had bothered to work it out. Settled at Pisa, he began to write commentaries on the *Almagest* with that object in mind, about the beginning of 1590.

Valla's precepts for demonstration in natural philosophy had made Galileo aware that his 'Natural Order and Local Motion' was not well organized logically. Definitions he had left until the third section were moved to come first, and motion was made the principal instead of a subordinate theme. Having written these revisions, Galileo put aside the whole treatise, with its

8   That tradition and the compromise effected with it will be discussed in chapter 10.

stress on philosophical problems, and began writing a 23-chapter, well-integrated 'Pisan *De motu*,' as it is usually called.[9] Galileo left it untitled, but a good descriptive title is 'New Doctrine of Motion.' It was written concurrently with his commentaries on the *Almagest*, from the beginning of 1590 to mid-1591.

This new and important treatise, which Galileo intended to publish and to follow as quickly as possible with his thoroughly revised commentaries on the *Almagest*, was tightly organized in two 'books,' with a wealth of cross-references among its twenty-three chapters. The first book contained ten chapters, ending *Et de hoc satis* – 'And enough of this.' It set forth the new doctrine of motion – new, that is, to Galileo's readers, not to him, for it had originated in the 1586–7 'Problems of Motion' and had evolved through his memoranda and 'Natural Order and Local Motion' to its definitive form. The second book contained detailed replies to objections grounded in the old and accepted Aristotelian doctrine of local motion, took up some problems of motion that had been listed but left undiscussed in 1586–7, and added new topics not treated by Aristotle. It was the custom when presenting any new doctrine to reply to possible objections and to extend it to new areas. But the second 'book' did something more; it opened with Galileo's statement of method:

The method that we shall follow in this treatise will be always to make what is [being] said depend on what was said before, and if possible, never to assume as true that which requires proof. My mathematicians taught me this method.

Book I had been organized from the previous text (stressing problems of a philosophical nature) by applying Valla's logical precepts. From that time on, Galileo followed instead the above two simple rules learned from his reading of Euclid, Archimedes, Ptolemy, and Copernicus – Galileo's 'mathematicians,'

Three chapters of the second 'book' are of particular interest

9 See my 'The evolution of *De motu*,' *Isis* 67 (1976), 239–50, and 'Galileo's pre-Paduan writings: years, sources, motivations,' *Studies in History and Philosophy of Science* 17:4 (1986), 429–48.

with respect to Galileo's later physics; I shall call them II-4, II-6, and II-7.[10] The last-named, 'By what agency projectiles are moved,' was probably revised and transferred from 'Natural Order and Local Motion,' for it was usual in the final lecture to deal with 'forced' motions.[11] As finally inserted in II-7, it had at its end Galileo's promise that his commentaries on the *Almagest* would, with God's help, be published in a short time. Clearly Galileo intended to publish his 'New Doctrine of Motion' first, and very soon. This promise was written in the latter part of 1591.

Chapter II-4, discussing motions down inclined planes, was the most important of all for Galileo's pre-Paduan physics, and there is no trace of any previous discussion of the topic in his writings on motion. Strictly speaking, what Galileo discussed in II-4 had never been examined by anyone, as he said, because his chief concern was with *speeds* of motions down inclines. Pappus had discussed the *force* required to move a body up an incline (but mistakenly, Galileo pointed out). Jordanus Nemorarius, in the thirteenth century, gave the first known correct statement of the *equilibrium* condition for a hanging weight balanced by a weight resting on an incline. Stevin published in 1586 a most ingenious (and utterly simple) argument establishing that same condition, but Galileo first reduced it to the lever law and went on to give his (mistaken) opinion about *motion* down an incline.

Galileo's mistaken reasoning about speeds on inclines dated back to his concept of merely temporary duration of acceleration in free fall, mentioned earlier. He candidly admitted that his rule of speeds was not borne out by experiment, ascribing the failure in part to 'material impediments' of the kinds mentioned earlier, and in part to a theoretical objection. None of those things had

10 In the 1960 English translation by I.E. Drabkin they are not numbered separately by 'books' and appear as chapters [14], [16], and [17]. The original manuscript gave chapter headings but not numbers, which Galileo intended to add when he was satisfied that the work was complete and properly arranged.

11 These were mainly projectile motions, then usually accounted for by medieval impetus theory, mentioned earlier in connection with acceleration in fall. How Galileo came to realize that impetus as an 'impressed force' was superfluous will be explained in chapter 6.

anything to do with the basic trouble, which was Galileo's neglect of acceleration. A decade was to pass before Galileo came to recognize that fact and move on to his mature mathematical physics.

Chapter II-4 is closely related, in an odd way, to II-6, the last full chapter to have been written (not very long after II-4). This chapter, on rotations of material spheres, had been foreshadowed by a paragraph in the old 'Problems of Motion' in dialogue form, dating back to 1586–7. It now changed the main course of Galileo's thought in physical science, concerning not only circular motion but also astronomy. There is no way to tell this story simply, but once it is understood, a host of common puzzles about Galileo's physics should disappear.

In Aristotelian natural philosophy, every motion was either natural or forced. Forced motions had been of scant interest to Aristotle because for him physics was the science of nature and especially of natural motions. Medieval natural philosophers developed impetus theory to remedy Aristotle's disinterest in forced motion.

It did not occur to anyone that a motion might exist which was neither natural nor forced, until Galileo examined motions on inclined planes. Motion down an incline was natural, that being undertaken spontaneously upon simple release. Motion upward on it must accordingly be forced. Hence on a level plane, neglecting any material impediments, no reason appeared why a body set in motion should change speed. No further force was required to continue level motion, and Galileo offered mathematical proof that the initial force could be less than any assignable force. Hence *continued* motion on a horizontal plane would be neither natural nor forced.

A horizontal plane, however, would be tangent to the earth only at a point, and motion from that point must take the body farther from the center of the earth, against its natural motion toward that center. Galileo concluded that even in theory, with all material impediments removed, motion could not continue on a horizontal *plane* uniformly and forever, but only on a surface concentric with the earth. The essential condition was that the distance from the earth's center must remain unchanged if speed were to remain uniform. For any approach toward that center must increase the speed of a heavy body, just as any removal from the center must require force and reduce the acquired speed.

It was by the concept of a motion neither natural nor forced that II-6, on rotations of material spheres, was related closely to II-4. Near the beginning of II-6, Galileo defined natural and forced motions a bit differently from Aristotle:

Now we have natural motion when bodies, as they move, approach their natural places, and forced motion when the bodies that move recede from their natural place.

It followed at once that a sphere rotating about the center of the universe moves with a motion that is neither natural nor forced, since it would be neither approaching nor receding from its natural place. Provided that the center of gravity of the sphere were at the center of the universe, it would not matter whether the sphere were homogenous or not. Galileo discussed the various cases of rotation with the geometric center or the center of gravity at the center of the universe, or supported near the surface of a sphere centered there. Only a heterogenous sphere rotating on its geometric center supported at a distance from the center of the universe would fail to move with Galileo's new kind of motion. In a marginal note added II-4 he christened motion neither natural nor forced 'neutral' motion.

It was while writing II–6 that Galileo, who still supposed the earth to be at the center of the universe, considered its daily rotation as a physical possibility. That would be a 'neutral' motion, neither natural nor forced, and no reason appeared why it should not be uniform and perpetual. At once he saw that if the axis of rotation were tilted to the ecliptic, astronomy would be greatly simplified. Annual revolution of the earth as assumed by Copernicus would not be required, nor the 'third motion' by which the earth's axis remained parallel to itself in the Copernican system. Galileo had found a system of astronomy, better not only than the hybrid geo-heliocentrism he had intended to propose in his commentaries on the *Almagest*, but than the Copernican system itself. Instead of three motions of the earth,

Galileo needed to assume but one – and one, moreover, already justified by his own physical reasoning about rotations of material spheres.[12]

As a matter of fact, unknown to Galileo, the semi-Copernican astronomy (as it came to be called) had already been proposed by N.R. Bär,[13] and Tycho had vigorously denounced him for it. Both for scriptural and for philosophical reasons, Tycho would allow no motion whatever to the central earth. So far as Galileo was concerned, logical simplicity of hypotheses alone mattered in astronomy, and he expected fame from presentation of the new idea in his commentaries on the *Almagest*, which he began at once to revise accordingly. Promising their speedy appearance (at the end of II-7), he laid aside his 'New Doctrine of Motion' as it stood in mid-1591 (and survives in manuscript unaltered).

Right at this time Galileo's father died, leaving him with new obligations and financial responsibilities as head of the family. Soon afterward it became clear that his expiring three-year contract at Pisa would not be renewed, mainly because he had antagonized professors of natural philosophy by an un-Aristotelian physics. In 1592 he was appointed professor of mathematics at the University of Padua, where he remained until 1610.

12 The path followed by Galileo is traced in my 'Galileo's steps to full Copernicanism, and back,' *Studies in History and Philosophy of Science* 18 (1987), 93–105.
13 This German name is usually Latinized as Ursus because it so appeared on the title-page of his *Fundamentum astronomicum* in which the new system was published at Strassburg in 1588.

CHAPTER FOUR

The Ancient Greek Background

In chapter 2 the medieval historical context of the law of fall was outlined, including Buridan's sophisticated causal account of fall in terms of impetus theory. That did not lead to the times-squared law, and indeed it was revived and revised to refute the law as poor physics soon after Galileo died. There had been, however, a much more ancient attempt to direct physics along the line eventually taken by Galileo, around 150 BC, and Galileo himself had started out along that ancient road before he learned that it had been opened in Greek antiquity – only to be promptly closed again.

The clue to this is given by a sixth-century commentator on the *Physics* of Aristotle named Simplicius, who wrote in part:

Alexander[1] carefully quotes a certain explanation by Geminus taken from his summary of the *Meteorologica* of Posidonius. Geminus's comment, which is inspired by the views of Aristotle, is as follows ... The only things of which astronomy gives an account, it can establish by means of arithmetic and geometry. In many cases the astronomer and the physicist [i.e., natural philosopher] will propose to prove the same point, e.g. that the sun is of great size or that the earth is spherical; but they will not proceed by the same road. The physicist will prove each fact by considerations of essence or substance, of force, of its being best that things should be as they are, or of coming into being, and of change.

---

1   Alexander of Aphrodisias, a commentator of the second or third century, whose interest was rather in the theory of knowledge than in physics or astronomy.

The astronomer will prove things by the properties of figures or of magnitudes, or by the amount of motion and the time that is appropriate to it. Again, the physicist will in may cases reach the cause, by looking to creative force; but the astronomer, when he proves facts from external conditions, is not qualified to judge of the cause, as when for instance he declares the earth or the stars to be spherical. And sometimes he does not even desire to ascertain the cause, as when he reasons about an eclipse; and at other times he invents, by way of hypothesis, and states certain expedients by the assumption of which *the phenomena will be saved*.[2] ... For it is no part of the business of an astronomer to know what is by its nature suited to a position of rest, and what sort of bodies are apt to move, so he introduces hypotheses under which some bodies remain fixed while others move, and then he considers which hypotheses will correspond to the phenomena actually observed in the heavens. But he must go to the physicist [i.e., the natural philosopher] for his first principles – namely, that the movements of the stars are simple, and uniform, and ordered – and by means of these principles he will then prove that the rhythmic motion of all alike is in circles, some being turned in parallel circles, others in oblique circles.[3]

In the second century BC it became necessary to clarify the boundary line between the domains of astronomy and natural philosophy. The reason was that by 125 BC Hipparchus had antagonized the philosophers by proving mathematically, from measurements extending back some four centuries, that the earth is not at the center of the sun's apparent motion through the zodiac. The cosmology of Aristotle's *De caelo* placed the earth at the exact center of the universe, and of all the celestial motions. In his *Metaphysics* Aristotle had adopted the system of homocentric spheres, invented by Eudoxus to show how all the apparent irregularities of solar, lunar, and planetary motions could be produced

2 The emphasis is, of course, added. Nothing needed salvation less than the phenomena – that is, the observed motions of heavenly bodies – and Geminus as an astronomer knew that. What needed to be saved were the principles and theories of natural philosophers, but he could not promise that and hope to establish a basis on which astronomers could proceed without interference.

3 Simplicius, *Commentary on Aristotle's Physics*, abridged from T. Heath, *Aristarchus of Samos* (Oxford 1913), 275

by uniform revolutions of spheres concentric with the earth, the axis of each sphere being carried by the next sphere, held at a fixed angle.

No actual measurements, of course, were necessary for the devising of homocentric spheres, nor had any been needed for Aristotle's logical and normative proofs that the earth must be fixed at the center of the universe. The useful work of astronomers, as for example the prediction of eclipses, depended on reasoning from careful measurements made over long periods of time. Natural philosophers could not allow a mere astronomer like Hipparchus to contradict them in their higher knowledge of essences and causes, but neither could astronomers yield to natural philosophers in the carrying out of the useful work of astronomy.

Geminus provided the compromise that was accepted on both sides until the time of Copernicus and Kepler. The key to this was the statement emphasized above, *the phenomena will be saved.* Cosmologists, who knew the causes of celestial motions, treated the task of astronomers as nothing more than to save appearances (for Greek 'phenomenon' simply means 'appearance'). Astronomical theories came to be regarded by later philosophers as mathematical fictions, devoid of any physical truth.

Now, cosmologers were also physicists, in the Aristotelian sense, so under the Geminus compromise they remained the final authorities on matters of physics. But Hipparchus had not only challenged Aristotle's cosmology; he had also written a book in contradiction of Aristotle's theory of fall and his conception of heaviness, striking at the very heart of accepted physics.

In Aristotle's physical theory, heaviness was treated as the cause of fall, while the natural striving of heavy things toward their natural place at the center of the universe was the cause that always directed fall straight toward that center; that is, the earth. Of bodies falling in the same medium, the heavier was the faster, while descent of a body through different media was swifter in the less corporeal medium. It was necessary for fall, as natural motion, Aristotle said, to be faster at the end than in its middle part; motion faster in the middle part was forced. To account for observed acceleration during fall, Aristotle said that the heaviness of a body increases in the process of fall.

The contrary accounts advanced by Hipparchus are preserved only in part, in commentaries by Simplicius on Aristotle's *De caelo*:

There is general agreement that bodies move more swiftly as they approach their natural places, but various explanations are adduced. Aristotle holds that as bodies approach the totality of their own element they acquire a greater perfection; and thus by added heaviness earth is carried faster to the center.

Hipparchus, in the book he wrote *On Bodies Carried down by Heaviness*,[4] declares that in the case of earth thrown upward by a certain power [*virtutem*] which projects it, the projecting power is the cause[5] of the upward motion as long as the projecting power overcomes the downward tendency of the projectile, and that to the extent that this projecting power predominates, the object moves the more swiftly[6] upward. Then as this power diminishes [1] the upward motion continues, but no longer at the same rate; [2], the body then moves downward under the influence of its own internal impulse, even though the original projecting power lingers in some measure; [3], as this power continues to diminish, the object moves downward always more swiftly; and [4], most swiftly when this power is entirely lost.

Now, Hipparchus asserts that the same cause operates in the case of bodies let fall from on high. For, he says, the power which held them back remains with them up to a certain point, and this opposition of contrarieties becomes the cause for the slower movement at the start

---

4 Περὶ τῶν διὰ βαρύτητα κάτω φερομένων. From the title it seems probable that Hipparchus composed an entire book on the subject of heavy bodies, of which our knowledge is limited to his theory of fall. It may have contained observational and perhaps even mathematical arguments, but Alexander's specialty was noetics rather than physics, and Simplicius was concerned with the philosophical aspects of the debate, as will be seen.

5 This was of course written before the compromise of Geminus, and Hipparchus doubtless regarded himself as a natural philosopher qualified to deal with causes and to oppose Aristotle on this matter.

6 Here Hipparchus, a mathematician, implied speed proportional to the effective net power. English translations give the word 'force,' best reserved to *vis*.

of the fall. Alexander replies: 'This may be true in the case of bodies moved or kept in place by force [vi] against their nature,' rightly saying 'but what was said would not fit when something upon coming into being is moved in accordance with its own proper nature to its proper place.'[7]

On the subject of heaviness, also, Hipparchus contradicts Aristotle, as he holds that bodies are heavier the further removed they are from their natural places. This, too, appears improbable to Alexander. 'For,' he says, 'it is far more reasonable to suppose that when there is transmutation to another nature, as when the light becomes heavy, this still retains something of its former nature while it is still at the very beginning of the downward fall and is just changing to that form by virtue of which it is carried downward, and that it becomes heavier as it goes along, than to suppose it still to be carried by the upward power which at the beginning detained it, and prohibited its being carried downward ... For if these bodies were to be moved downward more swiftly in proportion to their distance from above, it would be unreasonable to suppose that they exhibit this [same] property in proportion as they are less heavy ... For to hold such a view is to deny that these bodies[8] move downward because of weight. Now, also in things projected upward by force, and in those held up, and in those changing place upward when by nature they go down, the same is seen to be done.' These, Alexander says against Hipparchus; and he says well (as I deem), especially since if on account of heaviness the speed becomes less, as in weights thrown [upward?],[9] it is clearly impossible that they be heavier by detraction, the heavier being what is more remote and is moved more swiftly than the speed existing when nearer.

'The reason given by Aristotle for acceleration in natural motion, namely an addition of weight (or of lightness),' says Alexander, 'is a

7 Note that the objection raised by Alexander was pertinent to the physics of Aristotle, but in our physics would be regarded as an imaginary situation of a purely metaphysical kind. This will be of importance later, when the remarks of Galileo are considered.

8 That is, newly created and hence not previously acted on by any force

9 The Latin is *ut in lancibus librem*, not *in projectis* as above. Simplicius may have alluded to the slowing of weights thrown in any direction. Clagett notes that some believe Hipparchus's theory to imply an 'impressed force' (*Science of Mechanics*, 544–5). But that idea may have arisen from confusion of this passage by Simplicius with the opinion of Hipparchus that it rejected.

sounder reason and more in accordance with nature. Aristotle would hold that acceleration is due to the fact that as the body approaches its natural place it attains its form in a purer degree; that is, if it is a heavy body, it becomes heavier, and if light, lighter.'[10]

No Greek manuscript or Arabic translation of the book named above is known to survive. It is uncertain whether even Simplicius himself knew more about the book than what Alexander of Aphrodisias had cited, two or three centuries earlier, in order to raise his objections against it. The title, *On Bodies Carried down by Heaviness*, suggests a general treatise on heavy bodies, not just on the matter of acceleration in fall. That, alone, probably would not have resulted in composing a whole book; if it had, the author would certainly have adduced concrete evidence against the accepted theory of Aristotle. Simplicius did not explain the reason for which Hipparchus had composed this book,[11] but a highly probable historical reconstruction of the events follows.

Almost as easily observable as acceleration in fall is the equality or near-equality in times of fall of two bodies differing considerably in weight. Nevertheless, Aristotle cannot have made even such easy observations when he wrote, quite explicitly:

If a certain weight move a certain distance in a certain time, a greater weight will move the same distance in a shorter time, and the proportion which the weights bear to one another, the times too will bear to one another; e.g. if the half weight cover the distance in *x*, the whole weight will cover it in *x*/2.[12]

That is was not difficult to refute Aristotle on this matter is shown by the observation of another sixth-century commentator, Philoponus (though the phenomenon was overlooked by Simplicius):

---

10 Clagett, ibid., 543, quoting from M.R. Cohen and I.E. Drabkin, *A Source Book in Greek Science* (New York 1948), 209–11, with my modifications according to a sixteenth-century Latin text such as Galileo may have consulted.
11 Perhaps because he had to rely on Alexander's summary
12 *De caelo* 1.6 (273b.30ff.); Loeb trans., pp. 49–50

If you let fall from the same height two weights, of which one is many times as heavy as the other, you will see that the ratio of the times required for the motion does not depend on the ratio of the weights, but that the difference in time is a very small one.[13]

It is curious historically that this observation seems not to have been adduced until the sixth century, but neither does the near-equality of times appear to have been mentioned again until 1553. It then appeared not as an observation, but as a theorem published by Tartaglia's former pupil Benedetti. No reason is apparent why Hipparchus could not have observed this phenomenon in antiquity. As an astronomer he was a keen observer, and not one to hesitate about challenging Aristotle's science whenever it conflicted with observation.

As a mathematician, Hipparchus could have replied easily to Alexander. Even a heavy body newly created far from the center of the earth could never *begin* motion toward it at any *uniform* speed, starting from rest. Some initial acceleration must occur; and from observation by Hipparchus, the acceleration could not continue indefinitely – precisely *because* he believed that heavy bodies move downward because of weight. Accepting Aristotle's doctrine that the heavier of two bodies falls more swiftly than the lighter, and supposing acceleration to continue throughout a fall, the two would not strike the earth together after fall from a great height. Observing that they do, Hipparchus was obliged only to limit the *duration* of acceleration from rest. Nor is it surprising if that pioneer among systematic astronomers ventured also an improvement in physics, by 'saving the phenomena' of natural motions by heavy bodies from the 'physics' of Aristotle. Before the compromise of Geminus, not just astronomy but all of natural philosophy was open to criticism and improvement on the basis of the observed phenomena of nature.

The compromise of Geminus enabled astronomers to proceed in their useful work without interference from philosophers, who came to regard that work as a collection of mathematical fictions devoid of any physical truth whatever. The price of compromise

13 Clagett, *Science of Mechanics*, 546

in Greek antiquity had been a relinquishment of physics entirely to them. Accordingly, astronomy became a science (in the modern sense) nearly two millennia before physics did. Yet at the very beginnings of systematic astronomy, Hipparchus had held a very modern view of the proper procedure in both disciplines alike.

Galileo held that same view from the outset, and he reached the same conclusion as Hipparchus about the necessity of limiting the duration of acceleration. This is shown by autobiographical passages in his Pisan *De motu* of 1590–1. In his chapter on the cause of acceleration in fall, after remarking that the cause assigned by Aristotle had never appealed to him, Galileo wrote:

When I discovered an explanation that was completely sound (at least in my own judgment), I at first rejoiced. But when I examined it more carefully, I mistrusted its apparent freedom from any difficulty. And now, having finally ironed out every difficulty with the passage of time, I shall publish it in its exact and fully proved form.[14]

This is borne out by the school-day hailstone incident, by the hydrostatic analogy of the unfinished dialogue, and by a memorandum written probably in 1587. Having set forth his basic explanation of acceleration in fall, as the loss of a finite residual upward force at the beginning of motion, Galileo went on to say:

After I had thought it out, and happened two months later to be reading what Alexander says on the subject, I learned from him that this had also been the view of the very able philosopher Hipparchus, who is cited by the learned Ptolemy. Hipparchus is, in fact, greatly esteemed, and he is extolled with the highest praises by Ptolemy throughout the whole text of his *Almagest*. According to Alexander, Hipparchus too believed that this [force] was the cause of the acceleration of natural motion; but, since he added nothing beyond what we said above, the view seemed imperfect and was thought to deserve rejection by philosophers. For it seemed to apply only to those cases of natural motion which were

14 I.E. Drabkin and S. Drake, *Galileo on Motion and on Mechanics* (Madison 1960), 85

preceded by forced motion, and not to be applicable to that motion which does not follow forced motion. Indeed, philosophers were not content to reject the view as imperfect; they considered it actually false, and not true even in the case where the motion was preceded by forced motion. But we shall add the matters not explained by Hipparchus, and shall show that the same cause holds good also in the case of motion not preceded by forced motion, attempting to free our explanation from every fallacy. I would not, however, say that Hipparchus was wholly undeserving of criticism, for he left undetected a difficulty of great importance.[15]

What Galileo saw as having been missed was that the forced motion upward and the natural motion downward of heavy bodies 'are not really two contrary motions, but rather a certain motion composed of a forced and a natural motion ... so long as [both] intrinsic heaviness and extraneous lightness ... are mingled in the moving body.' Galileo's source for that insight had been an assumed complete analogy between fall through air and sinking in water. There was not a conflict between two contrary forces, but a composition of two inclinations, equally 'natural' in Galileo's physics at this time, just as the rise of one pan of a balance is no less natural, and necessary, than is the descent of the other pan. Hipparchus did not mention descent through media other than air, at any rate in what Simplicius preserved from his physics.[16]

It was long regarded as a serious puzzle that the young Galileo adhered in his Pisan *De motu* to the seemingly primitive notion of self-expending impetus (as it has been called[17]) after the concept of impetus as a kind of enduring impressed force had long become generally accepted among all but the most stubbornly

15 Drabkin and Drake, *Galileo on Motion*, 89–90
16 Ibid., 93. It was previously believed that Galileo had not read Simplicius but was misled by a secondary source into thinking that Hipparchus had neglected to explain acceleration in fall from rest. That is indeed what Galileo first discussed on pp. 91–2, but the 'foremost cause' on p. 93 had escaped even G.B. Benedetti in 1585.
17 The word *impetus* is better reserved to the concept initiated by Buridan, some two decades after the *vis derelicta* of Francesco di Marchia to be discussed below.

orthodox Aristotelians. That is no longer puzzling, once it is seen that at first Galileo set out to save the phenomena only of *natural* motions of heavy bodies, and in particular those of fall. Like Aristotle, Galileo in his early writings disregarded forced motions as of secondary interest. In contrast, medieval impetus theory was developed primarily to fill a hiatus left in physics by Aristotle's neglect of *forced* motions, and particularly to supply a cause for continuation of motion by a heavy body thrown after it left the thrower's hand.

Now, the novel theory of fall proposed by Hipparchus did not perish with his book. The concept of an incorporeal motive power is again found in the commentaries of Philoponus on Aristotle's *Physics*, in a passage suggesting that he was unaware that anyone else had already proposed it:

It is necessary to assume that some incorporeal motive force is imparted by the projector to the projectile, and that the air contributes either nothing at all or else very little to this motion of the projectile. If, then, forced motion is produced as I[18] have suggested, it is quite evident that if one imparts motion contrary to nature, or forced motion, to an arrow or a stone ... there will be no need of any [material] agency external to the projector.[19]

The concept of an incorporeal force imparted to a body seems to have survived in the Arabic word *mail*[20] and to have passed into European medieval natural philosophy as the *vis derelicta*[21] of Francesco di Marchia in the 1320s. There is little doubt that this concept, in turn, inspired the *impetus* of Jean Buridan two decades later. That seems a conceptual advance, because the impressed force thereby became a thing not evanescent, but of a

---

18 The personal pronoun here, and the absence of a discussion of fall in terms of projectile motion, suggest that Philoponus was not aware of the work by Hipparchus.

19 Clagett, *Science of Mechanics*, 509–10. Attention was called to the importance of Philoponus by Emil Wohlwill in his 'Ein Vorgänger Galileis im 6. Jahrhundert,' *Physikalische Zeitschrift* 7:1 (1906), 1–10.

20 See Clagett, 512ff.

21 Ibid., 520

permanent[22] nature, reduced only by conflict with an external resistance or an internal contrary tendency to motion. For that reason it is customary to view impetus as a step toward the inertial concept.

What has been traced is a source for the self-expending *vis derelicta* of Marchia in the Arabic *mail*, and a source for that in the incorporeal force of Philoponus, or in the work of Hipparchus as preserved by Simplicius. What followed in the fourteenth century was a 'revolution,' as historians say, when the ancient concept of an evanescent force gave way to the truly original medieval concept of an abiding force imparted to a heavy body. Because that concept appeared to anticipate in certain ways the concept of Newtonian inertia, scholars looked for something like that also in Galileo's books. But they looked in vain, for inertia is a dynamic concept, whereas Galileo's published physics avoided the notion of forces in nature and remained purely kinematic.

If we look at the fourteenth-century revolution in science as a fork in the road, rather than as one link in an unbroken chain of physics (as many do), we perceive Galileo resuming the older road after others had long taken the new one. What he may appear to have borrowed from medieval physics – for instance, the diminishing *vis derelicta* of Marchia – had its original source in antiquity and had never been lost before Buridan introduced the *impetus* concept.

Even in Galileo's early writings (before he moved to Padua in 1592) no trace of an inertial concept can be found, though in *De motu* he did attempt a dynamics. There the terms 'impressed force' and 'impetus' were used, interchangeably, but nothing of lasting value to Galileo emerged. Not until 1598 did Galileo find, in the *Questions of Mechanics* then ascribed to Aristotle, a statement which put into his hands the principle of conservation of motion that took the place of inertia in his mature science of motion.

It has always been known that Galileo's mathematical physics began with his study of Archimedes, whom he frequently men-

---

22 Concerning the meaning of 'permanent,' see my 'Impetus theory reappraised,' *Journal of the History of Ideas* 36 (1975), 27–46.

tioned with the highest respect. His method in physics was not Archimedean, but rather it was that of ancient Greek astronomy. The Greek astronomer-physicist Hipparchus had most nearly anticipated the seventeenth-century revolution in physics.

CHAPTER FIVE

Mechanics, Tides, and Copernicanism

It is hardly to be doubted that when Galileo moved to Padua late in 1592 he intended to publish there, or at nearby Venice, his new doctrine of motion and his revised commentaries on the *Almagest* presenting the semi-Copernican system of astronomy and cosmology. Expiration of his contract at the University of Pisa had removed any advantage he could gain by publication there or at Florence, whereas the fame he expected from the astronomy he believed to be original would cement him in his new and better-paid professorship at Padua. Yet neither manuscript was ever published, and Galileo's interest turned from motion as such to mechanics and other practical subjects for several years.

It happens that Galileo's inaugural public lecture in the Aula Magna at the University of Padua was an outstanding success, as a foreigner who had arrived there in 1589 reported in a letter that survives. He was a young Danish astronomer who had studied under Tycho Brahe and did not leave Denmark before Brahe started vigorously denouncing Ursus as an upstart and plagiarist for his having proposed the semi-Copernican astronomy in 1588. Galileo thus learned early in his stay at Padua that if he proceeded with the planned publication he would receive not fame, but scorn for ignorance of work in astronomy abroad, along with the enmity of Europe's most celebrated astronomer.[1]

---

1 It is not strange that he had not even known of the books by Tycho and Ursus while he was at Pisa. Tycho's 1588 volume was privately printed, not published, intended as the second in a three-volume treatise. Only one copy is known to have reached Italy, in 1590. It was sent to Gellius Sascerides, the Dane at Padua who

In 1593 Galileo wrote a syllabus of lectures on mechanics in Italian for his private pupils, expanding it somewhat in 1594. In the same years he wrote also on military architecture and on fortification. None of those was a university subject, but many young noblemen destined for military careers came to study at Padua, so professors augmented their salaries by special private tutoring for fees. Galileo's lectures on mechanics reduced the five simple machines to the lever law, offering nothing novel except in a brief section at the end on the force of percussion. That was not a usual topic in books on mechanics, and Galileo's early remarks on it are not without interest for the germ of a conservation concept they contain, something that was to become a principal aspect of his mature physics.

Undoubtedly it was Galileo's interest in mechanics that drew his attention to the phenomena of tides. In the Ligurian Sea near Pisa tides are hardly noticeable, but at Venice, hardly twenty miles from Padua, tides rise some five feet – though along the Adriatic coast generally they hardly exist. Moreover at Venice, where Galileo frequently visited while he lived at Padua, tides are quite conspicuous by reason of the many vertical pilings and buildings against which they rise and fall. The weight of such an enormous amount of water as is contained in the Venetian Gulf could not be lifted five feet twice a day except by some gigantic force, for which there was no apparent source, raising a problem for anyone interested in mechanics. The traditional explanation was dominion of Cynthia over Neptune, or 'influence' of the moon over the seas, not a force but much too vague for Galileo.

It was in 1595 that Galileo found an explanation for tides that he extended and developed over many years, finally making it the organizing theme of his famous *Dialogue* published in 1632. The earliest trace of it is in the notebooks of Fra Paolo Sarpi, Galileo's friend from his earliest days at Padua. It was with Sarpi that he discussed phenomena of motion repeatedly, and the key to Galileo's theory of tides was not force, but motions. The idea

heard Galileo's inaugural lecture, for transmittal to G.A. Magini, professor of astronomy at Bologna and no friend to Galileo. The work by Ursus, printed at Strassburg, was so little known that even Kepler, a German, did not hear of it until 1597.

probably occurred to him while he was traveling from Pisa to Venice on a barge carrying fresh water from the mainland.

Galileo noticed that when the barge changed speed, as by its bottom scraping sand in shallow waters, water in the barge surged forward and then continued to rock back and forth for a long time, rising alternately at the ends but not in the middle, where it could flow freely. Hence force was not necessary to lift a great mass of water; obstructed flow sufficed. The gulf terminating the Adriatic obstructed flow from the Mediterranean, and the analogy between water in the barge and that in the gulf struck Galileo. If the Mediterranean basin moved unevenly, tides at Venice but flow along the Adriatic coast below the city could be explained. (In fact, the period of reciprocation for the Adriatic is about one day, accounting for very large tides at Venice.)

Now if Copernicus were right about motions of the earth, a reason could be given why water in a large sea extending widely east – west *would* move unevenly – not as a whole, of course, but each and every part of it. As Sarpi wrote in his notebooks for 1595:

Any water carried in a basin, at the beginning of its transport, remains behind and rises at the rear, because the motion has not been thoroughly received, and when stopped, the water continues to be moved by the received motion and rises at the front. The seas are waters in basins and by the annual motion of the earth make that effect, being now swift, now slow, and again average throughout the diurnal [motion], seen in the moving of the basin diversely.

This is part of a succinct but adequate outline of the tidal explanation offered by Galileo to his friend Sarpi, who himself was no Copernican (and whose biographer reported that Sarpi had a tide theory of his own based on a single motion of the earth). Until 1595, neither had Galileo been fully Copernican, accepting only the diurnal rotation as a physically justified assumption. Tides convinced him that both Copernican motions of the earth were real; only the 'third motion' ascribed to its axis was a superfluous assumption.

Galileo's theory of tides is presented by most historians of science as a one-cause explanation, though in the *Dialogue* (as in 1595) two causes, interacting, were required. Large seas are continually *disturbed* by combined revolution and rotation of the earth, but that does not govern the *periods* of reciprocation; in Galileo's theory, the length and depth of each sea determine those. One sea might have a period of about six hours from low tide to high, as does the Mediterranean, but the period would differ, in Galileo's opinion, for another depth or a different east — west length.

It is evident that this flow-theory of tides, very different from the bulge-theory of Newton, is not thereby inconsistent with any possible dynamics. It is inadequate to explain all observed tidal phenomena, much as Newton's own theory was before Laplace greatly modified it. But suppose gravity to act only near the earth's surface, holding seas in place, and the earth to be held in orbit by some power completely different from gravitation. The earth could rotate daily with seas remaining calm. But if the earth also revolved around some center such as the sun, centripetal acceleration would not then be the same at places that differ in distance from the sun. Except at the poles, rotation would carry any given part of a sea-surface through a cycle of distances from the earth's center. Flows near shores would occur, and in principle the 'Galileo effect' must exist, however weakly, by virtue simply of the doubly circular motion of the earth.[2]

That water is not rigidly attached to the earth and does not at once receive, or lose, motion imparted to it by contact with a sea-bottom was duly noted by Galileo. His recognition that motion is conserved should not be regarded as entailing inertial, or any other kind of dynamic, conceptions. Presently it will be seen how a purely kinematic view of conserved motion came to modify profoundly Galileo's previous doctrine of motion.

In 1596 Galileo devised an instrument for measuring heights by sighting, incorporating Tartaglia's *squadra*. His treatise on

2   Galileo's own reasoning was of course devoid of centripetal acceleration. He regarded water as having a natural tendency downward by reason of its heaviness. That assumed, purely kinematic considerations could yield an explanation of tides in terms of a rotation and a revolution.

various uses of the instrument later became an appendix to his book on the 'geometric and military compass,' the pioneer general-purpose calculating device, developed during the years 1597–9 until it served to solve nearly any practical mathematical problem that was likely to arise. It resembled the sector, invented about the same time in England, but had many more scales and uses.[3]

Galileo's first statement that he was a Copernican (and that he had been for many years[4]) came in a letter to Kepler in August 1597, thanking him for two copies of Kepler's *Prodromus* of 1596. By a curious coincidence, both he and Kepler had become fully Copernican in 1595, in very different ways, Kepler having been inspired by a purely mathematical consideration and Galileo being convinced by the physical phenomena of tides.

In the same letter Galileo stated that he had written many arguments for Copernicus, and replies to objections against the new astronomy, but preferred not to make his opinion public yet because of its large number of foolish opponents. He also said that he had been able to explain physical phenomena that could perhaps not be explained in the old astronomy. Kepler guessed correctly that he alluded to tides, but Kepler did not believe that tides could be explained by motion of the earth and did not ask for further information. He did urge Galileo to speak out openly for Copernicus, if not in Italy then in Germany, but Galileo let this brief correspondence lapse.

One of Galileo's replies to objections survives in a long letter earlier in 1597 to a friend and former colleague at Pisa. Others were probably in his commentaries on the *Almagest*, now lost. As already said, Galileo was occupied until 1599 with work on his calculating instrument, writing three treatises at its successive stages of development, but in 1598 another event took place, of crucial importance to his thought as a physicist.

---

3 My translation of Galileo's book, with a history of the instrument, has been published by the Smithsonian Institution.

4 Galileo thought of his Copernicanism as dating from 1590–1, when he first accepted the sun as the center of motion for all planets but held the earth to be stationary, and then soon gave the earth daily rotation as in the 'semi-Copernican' system.

As a professor of mathematics Galileo lectured each year on Euclid's *Elements* and on either spherical or planetary astronomy. The only exception was the academic year 1598–9, when he omitted Euclid and, according to the university records, lectured on problems of mechanics. Now, mechanics was not a university subject, while geometry very definitely was, and to get official permission to depart so radically from the usual curriculum he must have had a clearly acceptable reason. No doubt he lectured on the ancient text called *Problems* (or *Questions*) *of Mechanics*, then attributed to Aristotle. Mechanics was not regarded as part of physics, so professors of natural philosophy would gladly have left it, like astronomy, to the mathematician to teach if he so wished, reserving cosmology and proper physics to themselves.

The ancient Greek treatise was, and is, extraordinarily interesting. It contains thirty-five questions on the widest variety of topics, starting from the balance and the lever and ranging through composition of motions, strength of materials, and useful devices, to eddies in streams. Its proposed answers, solutions, or explanations were intelligent and not dogmatic. I think it probable that Galileo first came on the work at Padua, for the appendix on the force of percussion in his syllabus on mechanics has language seemingly taken from *Problems of Mechanics*. The work appears not to have been known in the Middle Ages; it was first printed in Greek, at Venice, just before the beginning of the sixteenth century, with the authentic works of Aristotle. There were two Latin translations in that century, and several writers paraphrased or commented on the treatise.[5]

Galileo's lectures in 1598 were probably in the form of commentaries on *Problems of Mechanics*, with perhaps additional topics of practical value. In preparing them he would surely have paid close attention to the exact words, probably working from the Greek text, as was customary in commenting on Aristotle. A most un-Aristotelian phrase in Question 33 could hardly have escaped Galileo's attention; since that is closely related to the

---

5   See Paul Rose and S. Drake, 'The pseudo-Aristotelian *Questions of Mechanics*,' *Studies in the Renaisaance* 18 (1971), 65–104.

previous question, I shall cite parts of both in the English of the Loeb Classical Library edition:[6]

32. Why do objects thrown ever stop travelling? ... Or is it ridiculous to deal with these difficulties when we do not have the underlying principle?
33. Again, why does a body travel at all except by its own motion when the discharging force does not follow and continue to push it?

The phrase rendered above as 'its own motion' was τὴν αὐτοῦ φοράν, more like 'its self-carrying along,' in my opinion. Since in Aristotelian physics every motion must have a cause, and the most durable axiom of metaphysics is that nothing moves itself, the ancient author probably did not mean the expression seriously at all, but *ad absurdum*, to stress the importance of his proposed solution. But Galileo, reading the text carefully (whether in Greek or in Latin translation), gave a moment's thought to the seemingly bizarre phrase, and that was enough to dispel forever from his mind the notion of an 'impressed force' he had adopted to explain continued motion of things thrown when he was writing his 'New Doctrine of Motion' at Pisa, years before.

The concept of motion simply *conserved*, and not caused by a force somehow contained within a body thrown, was of enormous importance to Galileo's mature physics. Reasons for which the simple conservation of motion had never even been considered seriously in natural philosophy, set forth above, strikingly illustrate the ways in which causal reasoning long delayed the birth of recognizably modern physics. Decades later, when he composed his famed *Dialogue*, Galileo was fully conscious of many preconceptions that had once been his and that would make it very difficult for contemporary readers to grasp his mature doctrine of motion. He placed those in the mouth of Simplicio, who spoke for philosophers in the *Dialogue*, and had Salviati, Galileo's own spokesman, remove them. The ensuing interchanges in the *Dialogue* probably reflect Galileo's reasoning around 1598.

In the second 'day' of the *Dialogue* Galileo presented a new

6 See the volume 'Aristotle, Minor Works,' 407; the Greek text is on the facing page.

physics compatible with motions of the earth. The question soon arose how a heavy body dropped or thrown could keep up with the earth rotating at a thousand miles an hour, or even with a ship under sail. Upon that, Salviati, taking care not to bring in any notion of force, outlined Galileo's purely kinematic view. After several pages of discussion, Simplicio correctly summed up as follows:

You have made an assumption throughout which will not lightly be granted by the Peripatetic school, being directly contrary to Aristotle. You take it as well known and evident that the projectile when separated from its origin retains the motion which was forcibly impressed upon it there.

The phrase 'forcibly impressed' was unobjectionable; Galileo would never deny that force was required to *begin* motion of the projectile. But the issue was what *continued* the motion, and on this he had Simplicio make the usual superfluous assumption when he went on:

Now, this impressed force is as detestable to the Peripatetic philosophy as is any transfer of an accidental property from one subject to another. In their philosophy it is held, as I believe you know, that the projectile is carried by the medium, which in the present instance is the air.

Salviati did not point out the unwarranted step from motion 'forcibly impressed' to 'impressed force,' but proceeded to show the utter inadequacy of air to continue the motion. After two pages of discussion, he said:

There must be something conserved in the stone, in addition to any [action by] motion of the air.

And following still further discussion, Salviati identified what it was that the heavy body must conserve:

When you throw it with your arm, what is it that stays with the ball when it has left your hand, except that motion received from your arm which is conserved in it and continues to urge it on? And what difference is there whether that impetus is conferred upon the ball by your hand or by the horse [you are riding]?

Thus was 'impressed force' shown to be a causal fantasy not needed in physics, however vital to natural philosophy. The word 'impetus,' coined in the fourteenth century to name the imaginary force, had been redefined by Salviati early in the first 'day' to mean simply speed acquired by a heavy body, whether spontaneously or forcibly. Simplicio had forgotten that definition when he said in the discussion that Salviati had introduced and named an impressed force repugnant to Aristotelians. The entire section shows how carefully Galileo wrote his *Dialogue* for intelligent laymen of his day, and is well worth reading in full.[7]

Conservation of motion removed for Galileo a problem that had lingered with him since his acceptance in 1595 of both the main Copernican motions of the earth. So long as he remained a semi-Copernican, the center of the universe retained a special significance in his doctrine of motion. After 1598 he could dispense with that and consider only the center of the earth, or more properly its center of gravity, as the common center toward which all heavy bodies tended to move. In other words, he could see his way clear to composing a natural philosophy consistent with the Copernican system, a grander project than his earlier intention of publishing a new doctrine of motion followed by commentaries on the *Almagest* of Ptolemy. But he was busily engaged in perfection of his general-purpose calculating device, and there was also the fact that such a work would offend Paduan professors of natural philosophy far more gravely than just an open expression of his Copernican convictions that he had told Kepler he preferred to avoid.

The situation was altered in 1600 by a personal letter from Tycho Brahe, inviting correspondence and suggesting collabora-

---

7 See Galileo, *Dialogue concerning the Two Chief World Systems, Ptolemaic and Copernican* (Berkeley 1953), 141–56. Galileo had defined 'impetus' on pp. 25, 27.

tion in astronomy. Coming from the most famous living astronomer to an ambitious but still obscure professor of mathematics,
that was a proposal to which Galileo must have given consideration, but in the end he did not even reply to Tycho. That seems
a breach of common courtesy, unlike Galileo, that calls for some
explanation.

Having broken off correspondence with Kepler, refusal to work
with Tycho would leave Galileo isolated from the mainstream of
astronomical thought. He would get no hearing for his own ideas
unless he organized and published them himself. A good time to
organize them was now, for the pieces were all in hand, scattered
in manuscripts for the most part, and the rest thought out but
not yet written down. The calculating instrument was perfected
and no other pressing work was in progress. What Galileo proceeded to write is known, though it does not survive. In 1610 he
promised readers of his *Sidereus Nuncius* that he would soon
publish a book on the system of the world, and two months later
he gave its full title and a comment: 'Two books *De sistemate
seu constitutione universi*, an immense conception full of philosophy, astronomy, and geometry.' It headed a list of three major
and five minor works in hand, awaiting completion, in a letter
to the Tuscan secretary of state applying for the position of
mathematician and [natural] philosopher to Grand Duke Cosimo
II de' Medici at Florence.

Galileo never questioned Tycho's peerless ability as an
observer of planetary positions, but could not have respected his
capacity as a theorist. The geo-heliocentric system for which
Tycho was famed had occurred independently not only to Galileo
at Pisa, but to Duncan Riddell at Rostock and to Eliseo Roeslin
in Germany, and perhaps also to Simon Mayr, according to their
statements. Tycho regarded the others as plagiarists, but after
1543 it was easy to see value in making the sun the center of the
planetary orbits, without allowing motion to the earth. Within
two years Galileo went on from that to the semi-Copernican
position, vastly superior logically to the Tychonic, whereas Tycho
not only failed to recognize its superiority when Ursus proposed
it, but denounced him also as a plagiarist. That was clearly unjust,

as there was no way in which Bär could have got from Tycho the key idea of daily rotation of the earth on an axis tilted to the ecliptic.

As a junior collaborator with a famous astronomer who lacked theoretical insight, and was unjust into the bargain, Galileo could not hope for any net advantage. Neither could he refuse without stating his reasons, which would only offend Tycho still more deeply. Galileo chose the less impolite alternative of not replying at all, and set to work on *De systemate mundi*, the short title given in the *Sidereus Nuncius* to his promised book. Work on *De systemate* occupied him from mid-1600 to mid-1601. He then wrote a treatise on mechanics that led him in 1602 to new studies of motion that occupied him continually until mid-1609, when he heard of the Dutch telescope and for years turned his attention to astronomy. His treatises on motion and on mechanics were both listed in the 1610 letter, immediately after *De systemate mundi*.

It is not surprising that that work took Galileo a year to write, even though the material had already been compiled for the most part in other writings of his. No one since Aristotle had undertaken to devise a physics consistent with an astronomy and a cosmology. Kepler was the next to attempt that after 1601, quite differently from Galileo, for Kepler's physics was developed ad hoc to fit his cosmological scheme, already set forth in the 1596 *Prodromus*, of which Galileo had a copy before him as he worked in 1600-1.

There was never more than a single original manuscript of *De systemate mundi*, which Galileo destroyed late in 1632 when he was ordered to Rome to stand trial for publishing his *Dialogue*. It would have been seriously incriminating then if an unequivocally Copernican manuscript had been found among his papers by the inquisitors. Its destruction was a great loss to biographers of Galileo and to historians of science, but a good deal can be said of it now that all his manuscripts and working papers have been arranged and dated. Everything significant in his commentaries on the *Almagest* doubtless went into *De systemate mundi*, after which those were discarded, leaving a gap in our knowledge of Galileo's progress in astronomy that is only now being filled.

Just how much natural philosophy went into the first book of *De systemate mundi* is impossible to say; all that is certain is that it closed with a revised text of 'New Doctrine of Motion.' We have his revision of II-4 in the *Mechanics* of 1601, written immediately after *De systemate mundi*. It is not hard to guess how II-7, on the agency by which projectiles are moved, was revised after Galileo adopted conservation of motion as a principle. The old II-6, on rotations of material spheres, needed little revision and was probably placed at the end, as leading naturally into the second book of *De systemate mundi* – the exposition of Copernican astronomy by Galileo with replies to conventional objections against it. What natural philosophy came before the important new doctrine of motion was probably taken from the Aristotelian lectures of 1584 and 1588, abridged and revised to fit the purposes of the new work. Galileo certainly had those manuscripts before him in 1600–1, to serve as a guide in ordering his natural philosophy and making sure that nothing essential had been omitted.[8]

It is fairly certain that Galileo's tide theory was not put into *De systemate mundi*, because his 1595 treatise on that topic was listed separately as one of the five lesser works at hand in 1610. Also, his exposition of the Copernican system would have explained away the Copernican 'third motion' as superfluous – as not a motion but a kind of rest, in Galileo's later acute characterization of it. And finally, it is probable that Galileo showed the impropriety of the anti-Copernican argument that heavy bodies resting on the earth would be hurled from it by rapidity of the diurnal rotation. Copernicus mistakenly attributed that argument to Ptolemy (who had said nothing of the kind) and gave only the very feeble reply that any rotation natural to the earth would be natural also to detached parts of it. Galileo's reply in the later *Dialogue* involved the law of fall, which he did not yet have in 1600–1, but he could have answered then on the basis of implications of his analysis of rotations of material spheres in

8   On the back cover of the 1584 *De universo* there are drawings for the 1601 *Mechanics*, written right after *De systemate mundi*.

II-6 of his new doctrine of motion, together with his account of observed straight fall on a rotating earth.

The idea that if the earth rotated, a heavy body would not be seen to fall grazing the side of a high tower to its base but would strike the earth to the west of that was a perhaps not uncommon anti-Copernican argument at Galileo's time. Aristotle was not responsible for it, though it was attributed to him, and neither Ptolemy nor Copernicus mentioned it. Galileo brought it up in his *Dialogue* because it enabled him to introduce all three main principles of his mature physics simultaneously: relativity of motion, conservation of motion, and independent composition of motions. In *De systemate mundi*, I believe, he had dealt with the ridiculous argument much more simply, along the following line.

A heavy body always falls along a line directed straight to the center of the earth. Aristotle granted that, though he said fall was to the center of the universe, at which the earth was situated non-essentially. The side of a high tower lies along a line directed straight to the earth's center. Hence that is the line along which the heavy body must fall grazing the tower; it can make no difference whether the earth rotates or not.

De systemate mundi was an effective Copernican work, though Galileo still had much to learn about motion. How he was led to resume his study of motion in the new way described in chapter 1 will appear after his *Mechanics* is considered.

CHAPTER SIX

Cosmology, Mechanics, and Motion

It was said earlier that as soon as Galileo had composed *De systemate mundi* in 1600–1 he returned to mechanics, writing a treatise known by that title toward the end of 1601. At its end he alluded to a treatise on problems of mechanics appended to it, which has not survived. There is little doubt that it was an Italian version of his Latin lectures at the university in 1598–9 commenting on the ancient Greek *Problems of Mechanics*. Probably he had added some practical problems, for the first section of his *Mechanics* was 'On Useful Things That Are Derived from the Mechanical Science and from Its Instruments.'

In the 1610 letter listing works in hand awaiting completion Galileo described the third major work as consisting of:

Three books on mechanics, two with demonstrations of its principles, and one concerning its problems; and though other men have written on the subject, what has been done is not one-quarter of what I write, either in quantity or otherwise.

Clearly this manuscript was of quite considerable size, for Guidobaldo del Monte's 'Mechanics,' from which Galileo took his treatment of pulleys and their combinations, was a fairly large book. Treatment of the other simple machines in 1601 was more detailed and systematic than in the 1593–4 syllabus; like that, the *Mechanics* ended with a section on the force of percussion, also more extended than before. The surviving part of the 1601 *Mechanics*, together with its now lost appendix on problems of mechanics, formed two of the three books described in 1610. The third book dealt with strengths of beams and their resistance to breakage under loading, the first of Galileo's 'two new sciences' as finally published in 1638. Only a single page of the earlier manuscript survives.

The second work in 1638 was Galileo's 'new science of motion,' already described in the 1610 letter as:

Three books on local motion – an entirely new science in which no one else, ancient or modern, has discovered any of the most remarkable laws which I demonstrate to exist in both natural and violent movement; hence I may call this a new science and one discovered by me from its very foundations.

Such were the three major works listed as in hand at the end of Galileo's years at Padua when he applied for the post of chief mathematician and philosopher at the Tuscan court in Florence.[1] In order of composition they were *De systemate mundi* (1600–1), *Mechanics* (1601–2), and *New Science of Motion* (1602–9). Galileo was organizing the third in mid-1609, when he first heard of the Dutch invention of the telescope. It was indeed published in three books when finally completed in 1638, divided into uniform motion, naturally accelerated motion, and projectile motions, but I doubt that in 1610 Galileo intended to devote a separate book to uniform motion. More probably he had instead a book on motions of solids placed in water, the topic from which he had begun his studies of motion in 1586–7.[2]

There is no occasion for surprise that Galileo returned to mechanics immediately after writing *De systemate mundi* in 1601. The ancient *Problems of Mechanics* had not only put him

1 By 'philosopher' Galileo meant, of course, natural philosopher, or as we would say, physicist. By 1610 he was in a position to make physics a mathematical science grounded in measurements, as he fully intended to do, but the advent of the telescope and his discoveries with it diverted his attention to astronomy for many years.

2 Galileo published a book on the subject in 1612. Clues to his first manuscript of it are found in a shorter treatment that he wrote for the Grand Duke in 1611, which had been preceded by a manuscript that he left with Antonio de' Medici at Florence in 1608; see chapter 7.

on the track of a treatment of motion consistent with Copernican motions of the earth; it had done much more than that. Its first problem dealt with the balance and the lever; in 1638 Galileo credited Aristotle (as supposed author of the ancient treatise) with the first demonstration of the lever law, though not the first to have mathematical rigor, which he credited to Archimedes.[3] The second problem discussed independent composition of motions, a most un-Aristotelian concept, and in a later problem this was extended to a more general case.[4] Another problem suggested to Galileo his pioneer science of strength of materials, while still another discussed the paradox known as 'the wheel of Aristotle,' to which Galileo gave prominence in his theory of the structure of matter in *Two New Sciences*. One page in his miscellaneous papers dating mainly around 1600 bears a diagram of the 'wheel of Aristotle,' suggesting that his reflections on that paradox long antedated 1638. Moreover, one of the five lesser works listed in the 1610 letter was on the nature of continuous quantities, the underlying theme of Galileo's analysis of the 'wheel' paradox.

Even that is not all that Galileo got from the ancient work on problems of mechanics. In 1612 he published a book on bodies in water, in which he credited to Aristotle (again as putative author of the *Problems*) a principle that Galileo adopted instead of the hydrostatic principle of Archimedes, deriving that in turn from an ancestral principle of virtual velocities inspired by the 'Aristotelian' treatment of the lever law. The vital importance to Galileo of *Problems of Mechanics* has escaped attention only by reason of neglect in the past to arrange and date his unpublished treatises and working papers, or possibly also because modern attempts to trace the origin of Galileo's physics have emphasized possible sources in medieval speculative philosophy.

It was mentioned earlier that from the Pisan *New Doctrine of Motion* Galileo wrote a revised text of II-4 into his 1601 *Mechanics*. The 1591 text had attempted to compare the 'speeds' along

---

3   In the 1601 *Mechanics* Galileo gave his own proof, more direct than that of Archimedes, repeated in *Two New Sciences*.
4   Independent composition of motions was one of the three principles of the mature physics presented in the second 'day' of the 1632 *Dialogue*.

various inclines, mistakenly because acceleration was neglected. The 1601 revision dealt not with 'speed' on an incline, but with *moment*, a concept that from this time on became the very heart of Galileo's analyses of mechanics and of motion. Hence his basic error in the 1591 treatment of inclined planes simply disappeared a decade later – leaving Galileo still under his original misapprehension about brief acceleration in natural descent, while depriving that of any power to affect his results in mechanics. The revision also included a mathematical insight of importance. Reducing the inclined plane to the lever law as before, Galileo now noted that tendency to move down along the plane, at any point of rest on it, was the same as the (instantaneous) tendency to move down along a circular arc cutting the plane at that point.

What is principally of interest in the 1601 *Mechanics*, for the new developments in Galileo's work on motion that followed, is the concept of *moment*. Galileo took the word from Commandino, whose definition of 'center of gravity' had been 'that point in a solid around which parts of equal moment are arranged.' Galileo defined *moment* as the tendency to move downward resulting, not from heaviness alone, but also from placement, speed, and other things causing tendency downward.[5] When a professor of philosophy later disapproved Galileo's use of *momento* in a sense unusual for that Italian word, he replied that it was in common use by mechanics, as well as in the phrase 'an affair of great moment.' The next edition of the authoritative Italian dictionary compiled by the Crusca Academy included Galileo's sense for this word along with the usual meaning (an instant of time).

In effect, Galileo thought of *moment* as produced by a weight acting at a distance from a center, as on a lever – although it would have been improper in his mathematics to *multiply* a weight by a distance, as we do not hesitate to do. It is only because compound *ratios* were permissible that in Galileo's working

5   The word for heaviness was *gravità*, not to be confused either with our 'gravity' or with weight (*peso*), because it indicated a tendency rather than a force or the measure of a force. It was a natural tendency to move downward, though later in his *Mechanics* Galileo extended *moment* to any tendency to move in other directions.

papers it may appear to us that he formed products of the kind $S = vt$, as we do, but those were not formed *singly*. We cannot grasp Galileo's thought, or appreciate its rigor mathematically, unless we always remember that he worked only in ratios and proportionalities.

The principle that Galileo credited to Aristotle in his book concerning bodies in water, and which I see as ancestral to the principle of virtual velocities, was that bodies become of equal *moment* when they have speeds inversely proportional to their weights. That principle, in a curious form, was applied in some calculations on the earliest page found among Galileo's working papers on motion, f. 146, bearing only a single abbreviated word. When I first analyzed the calculations, in 1975, I misread that abbreviation as *mo.*, for *moment*, whereas it was truly *rao*, for ratio, as a critic was quick to point out. What Galileo had been calculating was a ratio of moments, so my error was not as fatal to my conclusions as the critic assumed. I dated the page to the year 1602, knowing nothing then of *De systemate mundi*. I would now date it to the latter part of 1601, when Galileo was ending that work and starting his *Mechanics*. The calculations on it are uniquely interesting, as they are related to Galileo's Copernican reflections nearly a decade before he had his telescope.

Galileo must have had before his eyes his copy of Kepler's *Prodromus* as he was writing *De systemate mundi*, for that is the only possible source of the data used in his calculations on f. 146. The first twenty chapters of Kepler's *Prodromus* were devoted to Platonist speculations about the *distances* of planets from the sun, of a kind utterly foreign to Galileo's conception of science. Kepler then turned to consideration of the *speeds* of planets, as related to their distances from the sun, and that did not fail to capture Galileo's interest. Kepler tabulated the number of days that each planet would consume in its annual orbit at the same rate of speed as that of Saturn, then the most distant planet known. Those were the figures that Galileo used in his own calculations, which were very different from Kepler's.

By Galileo's criterion, planets had no 'heaviness,' because they exhibited no tendency to move toward a center. But because lesser heaviness could be offset by greater speed in equalizing *moments*,

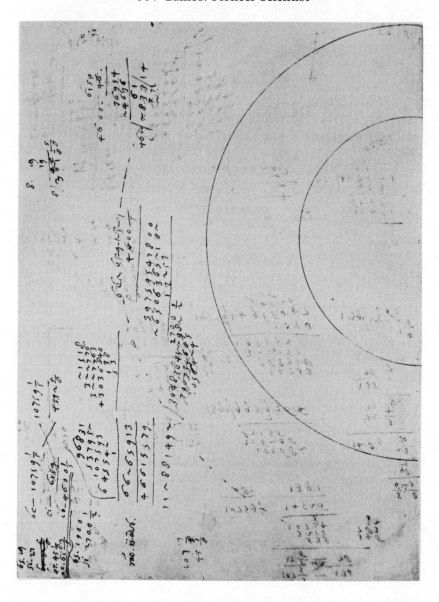

Figure 8. f. 146v. Calculation of planetary 'moments'
from Kepler's 1596 data.

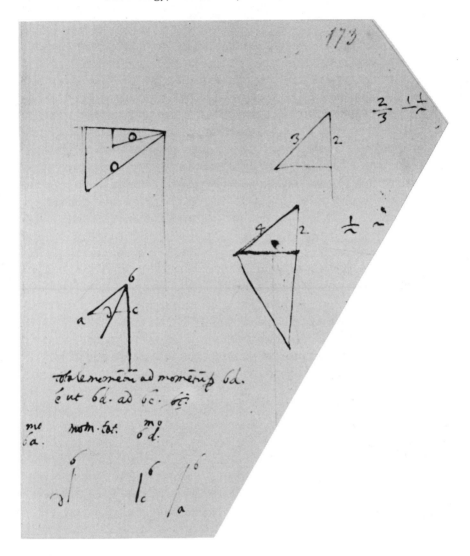

Figure 9. f. 173r. 'Moments' in the vertical and along inclines which led Galileo to his chord theorem.

it might be that speed could replace heaviness entirely in planetary *moments*, whose ratios Galileo began calculating. He soon saw that Saturn would exceed Jupiter in *moment* and sought a point beyond Saturn, distances from which would invert the ratios calculated. If he continued his calculations along the line of reasoning commenced on the lone surviving page, as I think it very likely that he did, Galileo could have found a relation that Kepler never stated explicitly, though it is implied by his later planetary laws: that mean solar-planetary distances are inversely proportional to the squares of the orbital speeds.[6]

Galileo did not publish this discovery, very favorable to the Copernican system, but he did put into the 1632 *Dialogue* a very interesting speculation about the origin of the solar system that he derived from it in 1608, as evidenced by notations in his working papers on motion after he found the times-squared law of distances in fall. Also, a letter Galileo wrote at Rome in 1633 while awaiting trial by the Inquisition included a bit of information about planetary speeds in the Copernican system that it would be hard to explain his recalling, far from his astronomical papers, unless he had carried out calculations of this kind.[7]

The earliest page of the working papers on motions of heavy bodies near the earth's surface bears the same watermark in its paper as the page of calculations just described, and no other page of Galileo's known to me has that watermark. On this page (f. 173) there is the abbreviation *mo.* for *moment*, associated with a sketch of balls on two planes differing in tilt, as also with a diagram relating vertical fall to descent along an incline. It is quite a remarkable page, for it shows that, two years before he had the law of fall and even before he had given up his mistaken idea that acceleration is of brief duration in natural descent from

6   The ratios I calculated from Kepler's data and Galileo's implied procedure are in my *Galileo at Work* (Chicago 1978), 65. Galileo's calculations were analyzed in my 'Galileo's "Platonic" cosmology and Kepler's *Prodromus*,' in *Journal for the History of Astronomy* 4 (1973), 174–91.

7   Orbital speed of the earth is the mean proportional between the speeds of Venus and Mars. No other astronomer would be likely to have noticed this curious fact, but the mean-proportional relation was of central importance to Galileo once he had the law of fall. See my 'A neglected Galileo letter,' *Journal of the History of Astronomy* 17 (1986), 99–108.

rest, Galileo saw how to determine the distance a body would fall vertically while a ball descended from rest along an incline to a given point. In other words, he had correctly compared *two* uniformly accelerated motions from rest when he did not yet know that either one was so accelerated. Any such comparison could have been made possible only by his thinking and working solely in ratios, rather than in absolute quantities or algebraically.

How a discovery of this kind is made often has nothing to do with how it is later proved to be true. This discovery would not have been made in 1602, had Galileo known then that acceleration must be taken into account, for he did not yet know how to deal with that. Acceleration certainly does have to be taken into account to prove his result, which amounts to this: To determine where the body falling vertically will be when another body that starts from the same point at the same time has descended to a given point along an incline, draw a perpendicular to the plane at that point; wherever that intersects the line of vertical fall will be the place sought.

Believing that acceleration was a brief event just to get a body up to its natural speed through air, Galileo ignored it and reasoned thus. Suppose that descent along some plane is half as fast as is straight fall. In II-4 of the 1591 'New Doctrine of Motion,' Galileo had mistakenly concluded that the *speeds* (not relative *times*) of motion were determined by the length and the vertical height of an inclined plane. Applying that incorrect rule to a plane tilted 30° drawn on f. 173, along which descent was assumed to be half as fast as in the vertical, he arrived at the above result. He then saw that the ratio of 'speeds' need not be 2:1. Since a right angle could always be inscribed in a semicircle, he was now in possession of what I call 'Galileo's theorem' – that the time of descent along any chord of a vertical circle to its lowest point remains the same regardless of the length and slope of the chord.[8]

In November 1602 Galileo communicated this theorem to his

8 My reconstruction of Galileo's 1602 proof of his theorem, on the assumption that acceleration is very brief and could be ignored, will be found in *Galileo at Work*, 67–8. His proof of the above result, in terms of *moment*, is found on f. 180r of the working papers.

friend and patron Guidobaldo del Monte, in a letter recommend-
ing the use of long pendulums (four to six feet in our units) for
certain experiments that Guidobaldo believed to refute a proposition that Galileo had sent previously. Galileo's own pendulum
observations at this time had undoubtedly been inspired by his
chord theorem. Along with that, he included in his letter the
theorem that the time is less, though the path is longer, when a
body descends along two conjugate chords to the low point, and
not along the single chord connecting the same end-points. I
call these two propositions 'theorems' because Galileo had valid
proofs for them on his defective assumptions; later, when he had
the law of fall, he of course supplied rigorous demonstrations. To
these theorems Galileo added a conjecture – he believed, and was
trying to find a proof, that along different *arcs* to the low point,
the times of descent were likewise equal.

At the end of this letter Galileo noted the failure of pure mathematical reasoning to be borne out by material tests, having probably in mind the separate sounds when bobs of equally long
pendulums, simultaneously released through differing arcs, strike
a block fixed at the low point of swing. This matter was raised in
chapter 1, but not another experiment that I believe Galileo also
carried out at this time. (Formerly I supposed that it occurred to
him only in 1604, when he began writing f. 189v1.)

Galileo's chord theorem implied that straight fall through the
vertical diameter of a circle, which is itself a chord to the lowest
point, takes the same time as descent along a very short and
almost horizontal chord to that point. Such a chord is insensibly
different from the subtending arc, in both length and slope.
Hence, if Galileo's conjecture about times along all arcs were
true, fall through that diameter should be timed by swing to the
vertical by a pendulum of length equal to the radius of the circle,
or at any rate very nearly timed by that. In fact, the timing
pendulum is longer than the radius of the circle, whence the
weight dropped through the vertical diameter would be heard to
strike the floor noticeably after the bob of a pendulum as long as
the radius, released simultaneously with the weight, had struck
the block. I believe that Galileo knew this a year before he started
writing f. 189v1. But the difference in times is only a small

fraction of a second, and so far as he was able to judge in 1602 it might be attributable to air resistance, not necessarily invalidating his conjecture about times along arcs to the lowest point in the absence of material impediments.

In *Two New Sciences* there is a passage that is certainly autobiographical and recounts Galileo's recollection, near the end of his life, of work he did around this time:[9]

In a small height it may be doubtful whether there is really no difference [in speeds of fall for different weights], or whether there is a difference but it is unobservable. So I fell to thinking how one might repeat many times descents from small heights, and accumulate many minimal differences of time, so that added together in this way they would make up a time not only observable, but easily observable ...

Thus opened an account of Galileo's early experiments with pendulums, to which he soon returned, but first he had gone on to say:

In order to make use of motions as slow as possible, in which resistance by the medium does less to alter the effect dependent on simple heaviness, I also thought of making the movables descend along an inclined plane not much raised above the horizontal.

From the continuity of the entire discussion it is safe to conclude that Galileo's careful observations of pendulums and of motions along inclined planes began about the same time, and in that order. That is what one would expect in the light of his having arrived at the chord theorem in 1602. The above passages indicate that Galileo was then still concerned with any effects of weight, or heaviness, on natural motions. It seems to me important to realize that weight, acceleration, and resistance of air were all in Galileo's mind at the start of his investigations of motion in 1602. Those are quantitative physical phenomena, on which previous natural philosophy could throw no light. Neither could astronomy, but the *method* of astronomers could. First must

9   See p. 87 in my translation of *Two New Sciences* (Toronto 1989).

come careful measurements of motions, and then the search for any relations among the numbers expressing measurements.

It was about 1602 that Galileo added at the beginning of his syllabus on cosmography the following memorandum, which came to be copied as an integral part of it by the scribe from whom his students could buy copies if they wished. Five scribal copies are known, all of which begin:

In the Treatise on the [Celestial] Sphere (which we shall more properly call Cosmography), as in other sciences, the subject should [first] be pointed out, and then we should touch on the order and method to be followed in it ... The subject of Cosmography is ... indicated by the word itself, which means 'description of the world,' ... It belongs to the cosmographer to reflect on the number and order of the parts of the world – on the shape, size and distance of each, and especially about their motions – leaving to Natural Philosophers consideration of the qualities of the said parts of the universe.

As to method, the Cosmographer usually proceeds in his reflections in four steps, the first of which embraces the appearances, or 'phenomena,' and these are nothing but sensate observations seen every day, as for example the rising and setting of stars; the ... moon's showing herself now crescent, now full, and again dark; the moving of the planets, etc. Second there are hypotheses, and those are nothing but some suppositions relating to the structure of the celestial orbs to correspond with the appearances, as when, guided by what appears to us, we may assume the heavens to be spherical and moved circularly, sharing various motions, and the earth to be stable ... at the center. Third, there are geometrical demonstrations with which, by means of some properties of circles and straight lines, the particular events that follow from the hypotheses are demonstrated. An finally, what has been demonstrated by lines being calculated, by arithmetical operations, it is reduced to tables from which, without difficulty, we later find at pleasure the arrangement of celestial bodies at any time.

It was not long after his letter to Guidobaldo that Galileo realized acceleration to be not temporary in natural descent, but enduring, reasoning from his observations of long pendulums and by the analogy between their swings and descents along inclined planes. He made a note of his puzzlement that theorems derived when ignoring acceleration nevertheless applied to accelerated motions when tested. During mid-1603 he was seriously ill and wrote little, but he did derive two least-time propositions from his chord theorem, and changed the word 'speed' to 'speeds' in an analysis of fall and descent along inclines. Toward the end of 1603 or the beginning of 1604 he recognized that in order to go further, he needed a rule for increase of speed from rest during straight natural descent. That was the origin of f. 107v, with which we began in chapter 1.

Of course when Galileo started f. 107v, he had as yet no way of measuring short intervals of time, nor could he legitimately measure speed as a kind of 'ratio' of a distance to a time. In principle those had no *ratio*, properly speaking, for in Euclid's mathematics 'ratio' was defined as a relation between magnitudes of the same kind – two distances, or two times, or the like. To express a relation between distances and times of motions it was necessary to establish a proportionality, that is, *sameness* of ratio for two distances, and two times. In uniformly accelerated motions, proportionality exists between distances and the *squares* of times, as Galileo was to discover in due course – but not on f. 107v, nor without the additional work described in chapter 1, and only by linking pendulum phenomena with those of free fall.

The immediate purpose of the work behind f. 107v was to find, if possible, a rule for successive increases in speed during a natural straight motion from rest. To slow descent, as Galileo said in the autobiographical passage in *Two New Sciences*, he used an inclined plane raised very little from the horizontal. The ball rolled at 1.7°; it was 20 *punti* in diameter, and the groove along which it rolled was about 9 *punti* wide, according to my calculations from various data recorded by Galileo when he was using the same apparatus in investigations to be described later. For the 1604 work at a slope of about 1.7°, it took the ball about 4 seconds to run the length of the plane, some 2100 *punti* (about 6 feet). Galileo marked and measured the distances run from rest at the ends of each of eight successive equal times, tabulating those measurements in *punti* on f. 107v.

Because the times were equal, distance between any mark and the next measured overall speed of the ball from one mark to the other. I call this 'overall speed' because what Galileo could find was not an average speed, or a mean (or middle) speed of the kind postulated by medieval writers on uniformly accelerated motion. His measures of speed were applicable to all *completed*[10] motions of any kind – uniform, accelerated, irregular, or even interrupted motions. That came about from Aristotle's definition of 'equal speed' and of 'greater speed,' as I shall explain, so that Galileo's thought and work at the outset of this pioneer investigation will be clear.

It may seem odd, but no one had ever yet defined 'speed' as such. That is why the Italian word 'velocità' (which was used by Galileo) was destined to give him much trouble during the next four years. I use the word 'speed' because 'velocity' has come to mean a vector quantity, involving direction of motion. Like most words ending with the suffix -ity, or -ness, velocity was coined from an adjective (meaning 'swift') and it just meant swiftness, a vague qualitative concept that nobody had trouble understanding, but that we would find hard to define precisely. The Latin *velocitas* was taken by natural philosophers to name a quality, intensity of motion – quite a different approach from quantitative definition, but perfectly adequate for philosophers. Aristotle's definition of 'equal speed,' the passing of an equal distance in equal time, was plainly applicable to any two motions whatever. For 'greater speed' he named two criteria – either a longer distance in the same (or equal) time, or the same distance in a shorter time – likewise valid even when the motions to be compared differed as to uniformity, continuity, or the like.

Thus it came about that the distances Galileo recorded were automatically measures of overall speeds during descent along

---

10 In Aristotelian natural philosophy, including its medieval developments, only completed motions were ever considered. For that reason proportional parts were always taken by starting from the end of a completed motion, in the direction of its beginning. Thus it was known that during the second half of motion from rest the distance traversed was three times that of the first half, but not in a way that suggested going on from 1, 3 to 1, 3, 5, 7, ... Nicole Oresme alone perceived that, and he did not develop the idea or apply it to the fall of heavy bodies.

his inclined plane. Taken successively from mark to mark, they went up as do the odd numbers, 1, 3, 5, 7, ... Because he was using distances for measuring speeds, he did not stop to consider them also as simple distances and sum them successively, getting the successive square numbers and at once recognizing the times-squared law of straight natural descent. In 1975 I supposed f. 107v to be the discovery document for the law of fall, because squeezed into the narrow left-hand margin of the tabulation were the first eight square numbers. Still, it puzzled me that those were in bluish ink, slightly smeared and a bit larger than the neat small numbers in black ink used for every other entry on the page. No reason appeared why Galileo should have delayed to note so interesting a discovery, if it had been made in that way.[11]

Of course it was not made in that way, as we have already seen in chapter 1. If it had been, Galileo could justly be accused of having jumped to a hazardous (although correct) conclusion. Natural motion down an inclined plane might not have the same law of acceleration as the motion of free fall. What Galileo did do was the opposite of jumping to any conclusion. The reason he put the square numbers in the margin on f. 107v was that after he had found the pendulum law, and from that the law of free fall, he wondered whether his times-squared law in fact held for descent along an inclined plane. Taking up f. 107v again, he verified the law as true for carefully measured distances along a plane. Only then did he conclude its generality for straight natural motions.

One more thing should be added before we proceed to events after the discoveries described in chapter 1. A ball rolling down an incline does not have the same speeds as a block of ice sliding down hot rails along the same incline. Because of the ball's moment of inertia, its rate of acceleration is only 5/7 that of the sliding weight. Galileo never became aware of that reduction, which did not affect any of his stated ratios and proportionalities among the phenomena he had measured.

So much the more should we admire the theory of proportion

11 See 'The role of music in Galileo's experiments,' *Scientific American* 232 (June 1975), 98–104.

in Book v of Euclid's *Elements*. Things were difficult enough to sort out at the beginning of recognizably modern physics. It was necessary to avoid mistakes in what was analyzed, but not to go on trying to analyze every effect in nature. As Galileo wrote in *Two New Sciences*, toward the end of his life, his new science of motion would serve as the elements, upon which minds more penetrating than his would build further.

CHAPTER SEVEN

Natural Motion and Horizontal Projection

Discovery of the laws of the pendulum and of fall, with the latter verified as also applicable to descent from rest along an inclined plane, opened the year 1604 for Galileo as one of great activity recorded in his working papers. In October he wrote out for Paolo Sarpi his first attempted derivation of the law of fall from an assumed principle. Precisely at that time a supernova appeared that diverted his attention to astronomy. Discussion of that interruption will be deferred to the next chapter on astronomy.

Work done immediately after the 1604 discovery, still on f. 189, was briefly described at the end of chapter 1. The correct compound-ratio rule for descents along planes differing in both length and slope was found, and then applied on f. 189r to verify numerically the second theorem sent in 1602 to Guidobaldo – that less time is consumed in descent to the low point of a vertical circle along two conjugate chords than along the shorter path of the single chord joining the same end-points. The calculation put Galileo in possession of a technique for determining time along an incline after motion had begun from a higher point, along a different incline or in straight fall.

This work, done immediately after discovery of the law of natural straight descent, involved no new measurements. Searches for consistent ratios were begun by assigning arbitrary times or distances to simplify calculation. Only the chord theorem and the times-squared law were needed. The first new rule had had a clear purpose in the verification of a theorem already known but previously derived without taking acceleration into account. The next thing to do, logically, was to find a rigorous proof of the chord theorem itself, and Galileo began almost at once on that problem, which he found much more difficult to solve. In fact, his full proof was not written out until late in 1607, though one basic lemma and the plan of attack were established in mid-1604.

Another rule soon found corrected the fundamental flaw in Galileo's original treatment of motions on inclined planes. He had concluded in 1591 that speeds along an incline and in fall through its vertical height were inversely as that length and height. By carefully drawing and measuring a diagram (using the ruler already described) and applying the mean-proportional rule relating distances and times in fall, Galileo found the correct relation – *times* of motion are as length of plane to its height. Later in 1604 Galileo used this relation for establishing an important theorem, and then realized that he had not yet proved the time relationship, having merely found its internal consistency by measuring a carefully drawn diagram representing the motions.

Many such steps in Galileo's rapid advance of knowledge about accelerated motions during 1604–7 can be traced from his working papers. Those were orderly steps psychologically, but they were very far from being demonstrative logical advances of the kind seen, for example, in Euclid's *Elements*. This work was, moreover, interrupted from time to time by a different kind of approach in the search for a deductive derivation of the law of fall itself, from some principle that would not require proof.

The first document of this kind was f. 152r, begun very soon after discovery of the law of fall and bearing entries made at various times until October 1604, when Galileo wrote out for Paolo Sarpi his first attempted derivation of the above kind. On f. 152r Galileo began by assigning 4 and 9 – the first two square numbers after unity – to distances in miles, obviously a purely arbitrary unit, and time of 4 hours for the first distance. For this imaginary natural descent he entered the 'speed' as 10, with no unit assigned. To the second distance he assigned the 'speed' 15, leaving the time in hours to be found. The two 'speeds' being $1+2+3+4$ and $1+2+3+4+5$, Galileo at first assumed that speeds grow as the natural numbers, while distances increase as

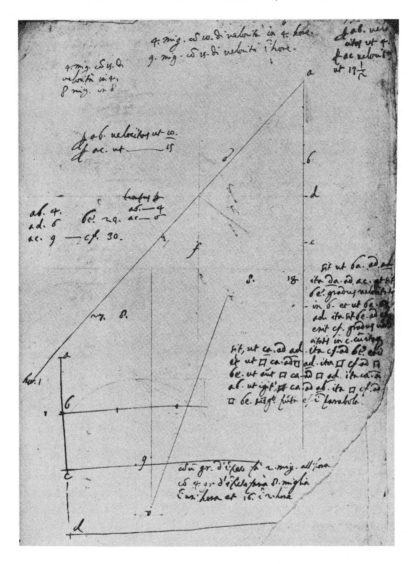

Figure 10. f. 152r. Soon after discovering the law of fall, Galileo attempted to justify it using the medieval concept of impetus, but without success. Around the beginning of October 1604 he added the part in Latin at lower right, finding that if degrees of speed were to increase as distances from rest, they would lie on a parabola through that point.

the square numbers. Application of the mean-proportional rule for times then led to an inconsistency. At the bottom of the page Galileo added a second diagram, representing accretions of 'impetus' to the natural tendency downward. In Italian (all other entries on f. 152r being in Latin), he wrote out the implications of medieval impetus theory as mathematicized by Albert of Saxony, perceived that an unexplainable doubling of a time would be required, and laid this page aside. Other brief entries were made on it that are of little concern here. Then, early in October, Galileo added to f. 152r a new and correct proportionality statement, noting that the points representing 'speeds' in the diagram he had drawn earlier would lie along a parabola.

A year before Galileo had adopted a representation of speeds by lines parallel to the base of a right triangle along whose vertical side he represented distances fallen from rest. Time as such was not separately represented in those early diagrams. The common notion among historians of science that Galileo long remained in doubt whether to make speeds in fall proportional to distances traversed or to times elapsed, and in October 1604 made a mistaken choice, is wrong. Until Galileo began measuring times of actual natural motions, it simply did not occur to him to make speeds, or anything else, proportional to times. Speeds in fall were clearly related to distances fallen, because they increased with growing distances. Just *how* they increased was less clear, and that was the question Galileo thought he could answer in October 1604. What he had in mind requires a digression here.

In November 1602 Galileo had told Guidobaldo that he had found a way to measure the force of percussion. That the same weight dropped from a height and then from doubled height has a doubled impact was now known to Galileo. Reasoning in 1604 that the weight stays constant and that only its speed changes during fall, he concluded that 'speed' is proportional to distance of fall from rest. At any rate Galileo took that as his principle when deriving the times-squared law of fall for Paolo Sarpi in October 1604. That is clear from the opening paragraph of f. 128, which bears his purported demonstration written at that time:[1]

---

1 In a letter dated 16 October sent to Sarpi, with whom Galileo had been discussing motion, he said that he had long lacked, but had now found, a principle from

I suppose (and perhaps I shall be able to demonstrate this) that the naturally falling body goes continually increasing its *velocità* according as the distance increases from the point from which it parted ... This principle appears to me very natural, and one that corresponds to all the experiences we see in instruments that work by striking, where the percussent works so much the greater effect, the greater the height from which it falls. And this principle assumed, I shall demonstrate the rest.

The Italian word 'velocità' was used above because, in the demonstration that followed, it was sometimes used for speed acquired at the end of a distance fallen – as above – and in other places for speed through a specified distance; in those instances I shall use the English word speed. The concept of an instantaneous speed, quite clear to us, was a self-contradictory notion for Galileo. Speed implied motion, and motion implied a lapse of time, however small. Aristotle used a different term, *mutation*, for change with no lapse of time. From 1604 to 1608 Galileo thought in terms of speeds over very brief intervals of time, not of mathematically instantaneous speeds, with the result that his theorems in terms of distances and times remained quite correct, while his relatively few attempts to bring in speed as such remained misleading and lacked rigor.

Of course when Galileo reasoned that *only* the velocità of a falling weight changes, he overlooked that its *square* must also change. Because impact depends in fact on v^2, Galileo missed the one measure of force (or rather of energy) that was available to him. Such errors of oversight in the midst of otherwise sound reasoning are precisely what make it slow work to recognize and correctly identify the thought behind Galileo's working papers on motion, which was not as confused as was its expression. So much for Galileo's so-called 'mistaken principle' of October 1604.

Only one brief note is found concerning descents along arcs of circles – a conjecture probably made in 1605. It is of no intrinsic interest, and Galileo never followed it up – or wrote anything about acceleration in such motions except to say in *Two New*

which he could derive the times-squared law and other propositions he had arrived at earlier.

Sciences that it is very different from the acceleration in straight natural motions. But f. 166r, on which he wrote the one note, is of considerable interest.[2] There Galileo had carefully drawn chords in a lower quadrant of a circle, and pairs of equal conjugate chords having the same end-points. He then calculated, to six places, the times of descent along chords in various combinations to the low point, starting from various heights. These times were not in *tempi*, or any unit used elsewhere, but were simply related to a time of 100,000 arbitrarily assigned to descent along the circle's radius. Two pages of his calculations survive, and a great many notes written on the document, from which it is possible to retrace Galileo's procedures and even to identify the source of an erroneous assumption made, found, and corrected by him. He was able to calculate times along inclines starting from any point, or any point higher on the arc from which a conjugate chord met the lower chord, and hence also times along broken chords, or along an incline after straight fall to it. Many theorems and problems in *Two New Sciences*, for some of which the working papers were written before Galileo left Padua for Florence, are of this type.

Of even greater interest, however, is the work Galileo began near the end of 1607. By that time he had accumulated a large number of theorems but had not troubled to arrange them in logical order. One, already mentioned, had still not been proved, and some others, including the chord theorem, lacked a lemma or would benefit from demonstration of a preparatory theorem. While Galileo was writing such neglected material on a group of folios, some previously started, and others blank sheets, he entered on f. 164 two brief notes that cleared away the puzzle he had felt in 1603, contradicted the statement he had made in October 1604 for deriving the law of fall, and put him on the track of the modern treatment of speeds in fall. These two notes also gave rise to a new series of careful measurements and experiments that provided Galileo with the parabolic trajectory for

2 This is the same page on which Galileo drew a line 169 mm long and identified that as 180 *puncta*, without which his *punto* could not have been confidently determined in metric units.

horizontally launched projectiles – his subject for the fourth and final 'day' of *Two New Sciences*.

The first note was written in a form particularly liked by Galileo, though by no means original with him; a seeming paradox was presented. (A diagram of the right triangle *ABC* showed its vertical side *AC* extending down to *D*.)

Remarkable! Is motion through the vertical *AD* swifter than that through the incline *AB*? It seems so, since equal spaces are traversed more quickly along *AD* than along *AB*. But also it seems not, because ... the time through *AB* is to the time through *AC* as *AB* is to *AC*, whence the same moments of velocità [exist in motion] through *AB* and *AC*; and indeed, that speed is one and the same with which in equal times unequal spaces are passed that have the same ratio as the times.

The paradox arose only because use of the word 'swifter' in connection with differently accelerated motions is ambiguous. A resolution of the paradox constituted the second note, and that corrected Galileo's mistaken statement of October 1604:

The moments of velocità of things falling from a height are to one another as the *square roots* of the distances traversed [from rest].

By the phrase 'moment of velocità' Galileo meant that factor in *moment* which arises from speed and not from heaviness (or weight). Galileo used the phrase 'moment of heaviness' when he distinguished the other factor in *moment*. The phrase 'degree of speed,' medieval in origin, meant a measure of speed *through* a distance; that is, during a lapse of time. The middle-degree postulate, or mean-speed theorem as it is often called, simply assigned to a uniformly accelerated motion from rest the 'degree' of speed existing *at* the middle *instant* of time for that motion. Obviously that was not a thing that could be actually measured, and literally there was not a 'middle' instant unless the number of instants happened to be odd.[3] Semantic problems attended

---

3　This is not a mere quibble, but another way of saying that a 'degree of speed' properly applied only to a motion over some distance, and hence involved some lapse of time, however small. A 'degree of speed' *at* a mathematical instant

Galileo's use of the medieval language of discrete instants and quantum-jumps of speed between them. That language equivocated with actual acceleration in natural descent – mathematically continuous by the times-squared law of fall. Semantic problems are created when new phrases are grafted onto an old accepted language, and they are notoriously harder to recognize than are logical problems. Galileo's language of *moment* dealt adequately with the distinctions needed in his new science of motion, but those were not identical with conceptions entailed by the two languages of impetus theory and of uniformly difform motion. It took Galileo a long time to remove the terminological conflicts.

Recognition that speeds acquired in natural descent are as the square roots of verticals from rest gave Galileo, for the first time, a means of testing experimentally his principles of conservation of motion and of independent composition of motions, both of which he owed ultimately to the ancient Greek *Problems of Mechanics*. Both of those principles were jointly tested by new measurements in 1608, as shown by f. 116v. That page records Galileo's discovery that the path of a body thrown horizontally is a semi-parabola, and was the first of his papers to be perceived to identify his unit of measurement for length in careful physical experiments.[4] The measurements and calculations found on it belonged to a much more sophisticated investigation than appeared to be the case when initially analyzed in 1972. I shall deal with it first as it then appeared to me, and later explain how there was more to f. 116v than could then have been foreseen.

When Galileo wrote the chapter on motions along inclined planes for his 'New Doctrine of Motion' in 1591 he did not state that motion once begun on a perfectly level plane, free from material impediments, would continue uniformly forever. He did say that the motion would be 'neutral,' neither natural nor forced, and he did prove that in principle it could be started by a force

remained undefined and could have had no meaning whatever in medieval physics.
4   See my 'Galileo's experimental confirmation of horizontal inertia,' *Isis* 64 (1973), 291–305.

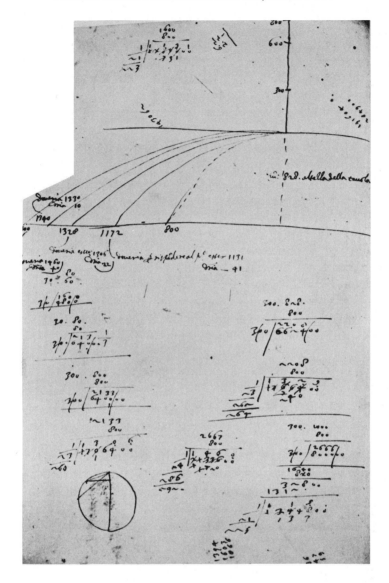

Figure 11. f. 116v. Experimental confirmation of conservation of horizontal speed and of composition of two independent tendencies to motion, giving the parabolic trajectory of a body projected horizontally.

less than any assignable force. His reason for saying no more was that apart from the fact that material impediments (such as air resistance) are in fact always present, motion along a plane tangent to the earth must carry the body farther from the earth's center, and thereby reduce its speed.

Although after 1598 Galileo had the concept of conservation of motion from the ancient *Problems of Mechanics*, he did not say that motion on a horizontal plane would continue uniformly even in his 1601 *Mechanics*, though that seemed clearly implied by two consecutive sentences in it:

Hence it is perfectly clear that on an exactly balanced surface the ball would remain indifferent and questioning between motion and rest, so that any the least force would be sufficient to move it, just as on the other hand any little resistance, such as that merely of the air that surrounds it, would be capable of holding it still. From this we may take the following conclusion as an indubitable axiom: That heavy bodies, all external and adventitious impediments being removed, can be moved in the plane of the horizon by any minimum force.

Galileo's seeming excess of caution in this matter reflects his habit when writing of avoiding any statements that he could not support by proofs. What he taught orally, especially to his more intelligent pupils, was less cautious. His most able pupil, Benedetto Castelli, came to Padua from Brescia no later than 1602 and soon began his studies under Galileo. In 1607 he mentioned in a letter to Galileo 'your Excellency's doctrine that although to start motion the mover is necessary, yet to continue it the absence of opposition is sufficient.' Hence it appears that, for all practical purposes, Galileo was satisfied in his own mind about uniformity of motion once begun along a horizontal plane, though he did not assert or imply it in print until much later.

Nevertheless, in 1607 Galileo still had no proof for the principle of independent composition of motions (also inspired by *Problems of Mechanics*), nor did he yet have any physical evidence in its favor much before he began writing on f. 116v. Aristotelian natural philosophers held that a body moves either naturally or by force, and even Tartaglia analyzed motion of projectiles into one part while an impressed force overpowered the natural downward motion, and another part after the latter overcame the (weakening) force and took the projectile straight down to earth.[5] Quantitative data alone could settle the matter, and that was what Galileo had newly obtained when he wrote f. 116v.

What had made this possible for the first time was the note on f. 164 that speeds in straight natural descent from rest are as the square roots of the vertical distances from rest. With that measure for speeds, Galileo now could launch a ball horizontally at different speeds whose ratios he could find. The horizontal advance during fall through air could be taken as uniform at the speed of launching, and measurement of such advances would show whether or not the supposed conflict between the two warring tendencies to motion in fact reduced either speed (or both).

On f. 116v Galileo drew two horizontal lines, for the floor and the top of a table. Above the table, along a vertical line, he recorded the heights 300, 600, 800, and 1000; between floor and table he drew a dotted line with the notation *pū. 828 altezza della tavola*, 'height of the table, 828 *punti*.' From the end of the tabletop to the floor he drew curved lines representing paths of a ball; where each met the floor, the measured advance from the end of the table was noted, together with a calculated advance and the difference between the measurement and Galileo's calculation. Galileo's calculations were made on the same page, using the rule of f. 164 and based on the first vertical distance to the table, 300 *punti*, and first projection, 800 *punti*.

The bronze ball had rolled down the grooved plane, through vertical heights to the table as indicated in *punti*, to a curved deflector, after which it rolled briefly along the horizontal and fell to the floor. Knowing the value of the *punto* in metric units, I calculated expected projections at Padua, and those agreed with

---

5 Tartaglia was already aware that the trajectory is curved from the beginning, but so slightly at first that for practical purposes it could be regarded as straight. Benedetti wrongly charged him with error in this matter, but did not doubt that the constant natural tendency conflicted with and weakened any contrary force until that ceased to act.

Galileo's measured horizontal advances within ±3%. In 1973 that seemed surprisingly good experimental work for 1608 to historians of science who generally had denied that experiment or measurement had played any appreciable role in Galileo's discoveries. His prompt recognition of the projectile trajectories as semi-parabolas is of course accounted for by the nature of the calculations he had made on f. 116v for a different purpose. On the working papers written immediately afterward in 1608–9, Galileo started his mathematical analysis of horizontal projectile motions (published in *Two New Sciences*).

Even more important conceptually than that is f. 91v, written at this time, for it was there that Galileo finally demonstrated that speeds in fall are directly proportional to times from rest. The contents of f. 91v were published almost verbatim in *Two New Sciences* near the beginning of the fourth 'day,' where Galileo explained composition of motions and vector addition in deriving the parabolic trajectory. That discovery was what determined Galileo to organize and publish the book on motion that he had tentatively started in 1604 and extended late in 1607. In the spring of 1609 he began discussing with the Roman mathematician Luca Valerio the two propositions on which he intended to base it, but news of the Dutch telescope abruptly ended this work.

Agreement within 3% between Galileo's measurements on f. 116v and my modern calculations aroused surprise (and a good deal of disbelief) in 1973 for one reason; a decade later it struck me as incredible for almost the opposite reason. At that later time I knew that by 1608 (and even in 1604), Galileo was experimenting accurately within ±4 *punti*; yet differences between calculations and measurements on f. 116v ran as high as 40 *punti*, without his having shown any sign of dismay. There was nothing wrong with his method of calculation, nor was the experimental procedure a very difficult one to carry out. Examining the whole matter in the light of further knowledge about Galileo's equipment and his sophistication as an experimentalist, I saw that he must have *expected* differences between the measurements and calculations recorded on f. 116v, and hoped to learn from those. No one could have believed in 1973 what is now pretty certain about f. 116v, a pioneer document in the history of systematic physical research.[6]

To begin with, the height of the table holding the inclined plane was *not* 828 *punti*, at any rate not at the time projections from it had been recorded on f. 116v. It was originally 800 *punti* high, and then raised to 820 *punti*. Galileo intended to raise it to 828, and quite possibly he did that, but I think he became too much interested in something else, more productive, to bother to do that. That is why he drew a *dotted* line from table to floor when he entered the number 828.

Next, I was mistaken when I translated Galileo's *doveria* on f. 116v as if he had meant 'it should be.' What he meant was 'it would have had to be.'[7] He did not feel exasperation with actual measurement for its failing to agree with theory when writing, of the second measured distance, 'It would have had to be, in order to conform with the first, 1131.' It had been 1172, and Galileo duly recorded: 'difference 41.' Thereafter he simply wrote *doveria* and *differenza* (abbreviated) at each of the other entries, giving the appropriate numbers. There would be no point in Galileo's having entered the differences unless he had varied something in his experimental procedure, and wondered how great had been the effect.

When a ball rolls atop a guiding groove, part of it remains always below the level of the groove, affecting the ball's moment of inertia. A larger ball or narrower groove, more nearly results in full acceleration (of a ball rolling atop a smooth plane). To calculate theoretical results from Galileo's data we sometimes have to know the size of his ball and the width of the guiding groove. In the case of horizontal projections, now in question, this is unnecessary; Galileo did not change balls or planes. But he did use different methods of horizontal deflection, as well as

---

6   A full discussion of the work done by Galileo in 1608 that lay behind f. 116v (and some of his other papers) can be found in my 'Galileo's accuracy in measuring horizontal projections,' *Annali dell' Istituto e Museo di Storia della Scienze di Firenze*, 10:1 (1985), 3–14.

7   In John Florio's Italian-English dictionary printed at London in 1611 there is a remark that nothing in the language causes more ludicrous errors among English-men than conditional forms of modal auxiliary verbs in Italian. Florio explicitly identified *doveria* at that time with the form now written *doverebbe*.

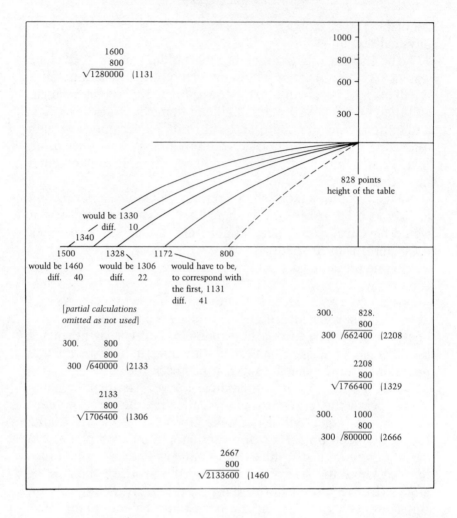

Figure 12. Partial English version of f. 116v;
original shown on p. 107.

two heights of the table. Most of the supposed 3% variance from modern calculations arose from assuming constant table height and rollings on an ungrooved plane. The actual ball and groove width reduced acceleration 5.75% below that of roll on full radius.

In *Two New Sciences* Galileo described his preparation of a grooved plane, saying that after it had been smoothed and cleaned very carefully, vellum was glued in it. It is in fact necessary to smooth such a plane, because even a small irregularity along an edge of the groove may send the ball out of it on a plane that is tilted fairly steeply. But there seemed no point in lining it with vellum, as the wording appeared to imply. What Galileo did was to glue limp vellum across the groove, not stretched tightly. Weight of the ball then pressed this into the groove, which fact had a striking effect on its acceleration. Because the lowest point of the ball was always in contact with a surface offering friction, acceleration was the same as for rolling atop a smooth plane; that is, maximum acceleration was assured. Galileo had learned this in the course of previous work underlying f. 116v.

Here I should say that although the work on f. 116v led to discovery of the parabolic trajectory, Galileo had then already confirmed the composition of uniform horizontal and accelerated vertical motions by measurements probably consistent within one or two *punti*. His working papers for that are not preserved, but Galileo would not have simply accepted discrepancies of 40 *punti*, as assumed in 1973 when f. 116v was first analyzed. Preliminary work had shown that transition from the plane to the horizontal was not entirely satisfactory without a curved deflector, best if grooved the same as the plane, to prevent any bounce at the table. Bouncing there would hardly affect projection, provided that the ball returns to the table and rolls at least briefly there (so as to assure horizontal projection), for little energy would be lost. But smooth transition from the roll down the plane to roll on the table is desirable, and calculation shows that, in the work for f. 116v, two different forms of deflection were used.[8]

---

8 Of course the ball would not be rolled down a plane to strike the table at the end of the grooved surface; a board would receive the ball's lowest point just as its center passed the perpendicular to the plane through the end of the groove.

Now, it makes a measurable difference whether the ball rolls in the groove until the instant of projection, as it would if the grooved deflector extended two or three inches horizontally, or whether the ball rolls briefly at the end on a smooth ungrooved surface. In the latter case the ball will acquire greater speed before projection than it had at the end of its roll down the plane, for any roll on the ball's lowest point, however brief, is upon a greater radius than in the groove, but with virtually the same rate of angular rotation as before. For Galileo's ball and groove, I calculate the increase of linear speed to be 2%, almost exactly.

Galileo, noting that projection became longer with the final horizontal roll outside the groove, would naturally wonder just how much longer it could be made. It was then that he thought of gluing limp vellum over the groove. That way the ball was still guided straight, but was at all times rolling on its lowest point. After vertical descent of 300 *punti*, I calculate a projection of 804 *punti* after drop of 800 if the ball rolls at all times in the groove. Rolling in the groove except for a brief terminal roll on a flat surface, projection is 820 *punti*; and finally, rolling on vellum and then on horizontal wood, projection becomes 828 *punti*.

On f. 116v Galileo drew a *dotted* curve to the point on the floor marked 800; his solid curve in this instance meets the floor slightly beyond that. Galileo had measured 804 *punti* (to one *punto* or so); to simplify his calculations he used 800, as he indicated by the dotted line. At top center of f. 116v is his calculation for an initial descent on the plane from 600 *punti*, with the table height 800 *punti*. Here he used his customary short form of calculation where we would multiply by $\sqrt{2}$. The identical form of calculation is seen at lower right, for a table height of 820 *punti*. As said earlier, none of the recorded projections was for table height 828 *punti*. Instead of raising his table again, Galileo turned to the more interesting and more fruitful study of parabolic trajectories at this point.

Shortly after writing f. 116v Galileo entered on f. 114v a set of numbers and a freehand sketch that became the key to detailed information about his experimental apparatus and procedures in the study of projections. The related investigations, unlike those of horizontal projections, were not set forth in *Two New Sciences* or mentioned elsewhere in Galileo's writings. They will be discussed in the next chapter, along with Galileo's other activities in physics before he turned to astronomy in mid-1609.

CHAPTER EIGHT

Oblique Projection;
Other Physics

In the summer of 1608 Galileo was requested to journey to Florence but appears to have wished to remain at Padua, though for several years he had spent the summer at Florence. He was, however, virtually commanded by the Grand Duchess Christina to be present at the wedding of the young Prince Cosimo, to whom Galileo had dedicated his book on the geometric and military compass in 1606. From the handwriting on the working papers of 1608, it appears to me that he had just finished f. 116v and begun his analysis of horizontal projections when he left for Florence, returning in October in ill health. Hence this is an appropriate place to interrupt the story of his work on motion and to mention Galileo's other activities in physics around this time.

For the festivities at Florence an esplanade was to be built on the Arno, and Galileo was drawn into discussions with engineers about its construction. Also present was Giovanni de' Medici, a military captain who favored a plan that Galileo knew to be unsound in principle. It was based on an idea that later became the subject of a bitter dispute between Galileo and Aristotelian natural philosophers: that floating on water depends on the shape of the floating solid. Galileo persuaded the engineers that only the density of the material matters, in this case wood, warning them that a disaster might occur if confidence were placed in the scheme favored by his adversaries. For several years thereafter Giovanni de' Medici was known to be hostile toward Galileo, whose philosopher opponents at Florence and Pisa enlisted him on their side, dedicating several books to him.[1]

It is probable that Galileo wrote out at Florence a treatise on the behavior of solids in water, which had been a principal topic of interest to him since 1586. Such a treatise must have been thoroughly Archimedean in content, though the demonstrations would have been 'more physical and less purely mathematical' than those of Archimedes, as were the proofs in his 1586–7 dialogue on problems of motion. This treatise is not found among Galileo's manuscripts, he having left it at Florence with Antonio de' Medici. A good idea of what was in it, however, can be gleaned from a manuscript that he addressed in 1611 to the grand duke Cosimo II de' Medici, the young prince of 1608 having ascended to the throne the next year. Galileo's book on hydrostatics will be discussed in chapter 11.

Soon after 1600, when William Gilbert published his famous book on magnetism, a copy was given to Galileo by a professor of natural philosophy at Padua, probably Cesare Cremonini, Galileo's personal friend and stout adversary on everything pertaining to science that in any way departed from Aristotle. Galileo remarked that the professor probably feared that Gilbert's work might infect other books on his shelves. Around 1602 Galileo and his Venetian friend Giovanfrancesco Sagredo engaged in some magnetic experiments, especially in armaturing lodestones as recommended by Gilbert. Galileo became quite expert at doing this, producing magnets that could sustain many times the weight of the original lodestones before they were smoothed and encased in iron. One very remarkable specimen (now on display at the museum of history of science at Florence) was purchased from Sagredo by the Tuscan court, Galileo intermediating. The

---

1 Giovanni de' Medici was the illegitimate son of a grand duke, not without influence at the Tuscan court. His half-brother Antonio was a warm friend and supporter of Galileo who will be mentioned again presently. In 1618 Giovanni praised Galileo's invention, for military use, of a binocular low-power telescope (called the *celatone* and worn on a helmet), as more admirable than invention of the telescope itself.

negotiations for Sagredo's armed lodestone were concluded in 1608.

It was also in 1608, at Padua, that Galileo carried out an interesting experimental attempt to measure the force of impact, described in a manuscript which at one time he considered adding to *Two New Sciences*. It was in fact not published until long after his death, in the second edition of Galileo's collected works.[2] The apparatus used was a stout-beam balance, having at one end a pair of buckets, one above the other; in the bottom of the upper bucket there was a hole, the size of a hen's egg, that could be suddenly opened. This bucket was filled with water, and the balance was leveled by adjusting a bucket of sand hung from the other arm. The plan was to add enough sand to restore this to level while water was falling into the lower bucket, Galileo expecting that the impact of falling water, added to the weight on that side, would make it outweigh the original sand. To the surprise of all present, that was not at all what happened. The instant the hole was opened, that side of the balance rose, and then gradually returned to level as the water moved from one bucket to the other.

Galileo's discussion of this experiment and its extension to the behavior of equal weights suspended over a pulley is of great interest, not only for its treatment of the force of percussion, his main intention, but also for the introduction of an inertial motion when the two equal weights hung over a pulley were set in motion. In a sense, Galileo concluded, the force of percussion is infinite, for on the one hand it can set a body into uniform motion indefinitely continued, while on the other hand there is no resistance, however great, on which any small impact remains entirely without effect. Or, as Galileo saw things, an infinite uniform motion can be begun by any small impact, which also can move a weight against any resistance through a finite distance.

Galileo's new science of strength of materials has been men-

---

2   My English translation is included at the end of Galileo, *Two New Sciences*
    (Toronto 1989). In this added 'day' Galileo retained Salviati and Sagredo as
    interlocutors, but replaced Simplicio with his own former student, Paolo Aproino,
    as one who had been present at the experiment and its discussion.

tioned as having probably originated in 1602, soon after the 1601 *Mechanics*. In the second 'day' of *Two New Sciences* we have it in a form to which additions had been made as late as 1634. That it had already reached a fairly advanced stage by 1608 is known from the lone page of the original manuscript that still exists. I have little doubt that although only the folder of working papers on motion is still preserved, similar folders on pendulums and on strength of materials once existed.[3] Three pages among the notes on motion clearly relating to pendulums, one page relating to sunspot measurements, and one page bearing a lemma on division of a parabola were doubtless transferred from folders whose main contents are lost. The lemma on the parabola is easily recognizable; it appears almost verbatim in the second 'day' of *Two New Sciences*. How it got into the folder on motion is obvious; as Galileo embarked on his detailed analysis of the parabolic projectile trajectory in 1608, he extracted from his folder on strength of materials a useful theorem on the parabola.

The proposition for which the lemma was written was quite an advanced one. Galileo sought the shape of a beam that would just support, without breaking, the same load at any point along it. One side of that beam being straight, the other will be parabolic or elliptical, depending on whether support is at both ends or at the center. Galileo's published solution did not consider both cases and was later challenged as defective, but the problem was an advanced inquiry for his pioneer science. It implies that by 1608 he had reached most, if not all, the preceding propositions in the book that he did not publish until thirty years later.

Illness on his return from Florence did not detain Galileo long from work in various areas of physics, as seen not only from his working papers but also from a letter he wrote to Antonio de' Medici on 11 February 1609. Because he had had ample opportunity at Florence to talk with him, it is likely that the discoveries listed in this letter indeed belonged to the last two months of

---

3 Among the notes on motion there are two pages, once cover pages of folders, labeled 'pertaining to motion.' Diagrams were drawn and calculations were made on them from time to time, so the former cover sheets were cut apart and saved as separate folios.

1608 and the beginning of 1609. After some matters of a personal character, Galileo wrote:

Since my return from Florence I have been occupied in some reflections and various experiments relating to my treatise on mechanics, in which I hope that most things will be new and not touched on by others in the past. And only recently I have finished finding all the conclusions, with their demonstrations, relating to the strengths and resistances of beams having various lengths, sizes, and shapes; how much weaker they are at the middle than at the ends, and how much greater weight they sustain if that is distributed along the whole beam rather than at one place, and what shape it must have so that it is equally strong everywhere – which science is very necessary in fabricating machines and edifices of every kind; yet no one has dealt with it.

Now I am about some questions that remain for me concerning projectile motion, among which many pertain to artillery shots, and just lately I have discovered this: That the [artillery] piece being put on some place elevated above the plain of the countryside and aimed perfectly level, the ball leaves the gun and, whether driven by much or little powder, or even by only enough to make it come out of the gun, it always goes dropping toward the earth at the same speed, so that for all level shots the ball arrives at the earth in the same time – whether the shots are very long or very short, or even if the ball just comes out falls plumb to the plane of the fields.

And the same happens for elevated shots, which all travel the same time whenever they rise to the same [maximum] vertical height.

Here Galileo drew a diagram showing, between two parallel horizontal lines, shots from the lower rising to the higher and returning along parabolic paths of various 'amplitudes,' as he called the horizontal distance in his analysis of trajectories in 1608–9. I shall return to this discovery presently; the letter continued:

... In the matter of water and other fluids, partly already with you, intact, I have likewise discovered very great properties of nature, but shortness of time prevents my writing to those at present because I have other letters to get off.

It happens that the main discovery meant here is positively identifiable from a letter of Daniello Antonini, a student of Galileo's who left Padua in 1609. Galileo had challenged him to discover a 'balance of equal arms in which one ounce of water on one arm may easily raise a hundred pounds of weight on the other, by means of that force by which a boat can be floated in a few gallons of water.' In other words the hydraulic lift, and one other aspect of the hydrostatic paradox, were in Galileo's possession by early 1609.

The statement to Antonio de' Medici that shots with a cannon elevated to any angle will take the same time from the cannon's mouth to the ground whenever the highest altitude reached is the same (even though the cannon is mounted far above the level at which the ball strikes) appears to be associated with f. 175r of the working papers. There Galileo drew a very steep inclined plane at whose end was a deflector that would send a ball upward rather than horizontally. The path of projection is drawn as rising from and returning to a horizontal line parabolically, and along the sloped plane are many dots, suggesting measured lengths of initial roll. One could, with such an apparatus, verify the above statement experimentally. I believe Galileo did that early in 1609, when he began the work on oblique projections that is described below.

What Galileo did not mention in his letter, or in any later book, is an investigation at this time that was related, but only indirectly, to the above discovery about projectiles launched not horizontally but obliquely. The clue to this investigation was found on f. 114v, mentioned near the end of the previous chapter. It is from that, and from the even more remarkable f. 81r, that details about Galileo's grooved inclined plane and bronze ball have been gleaned that could not have been obtained from f. 116v. From that record of *horizontal* projections, not even the slope of the plane could be deduced. I can now say with fair assurance that its slope was arctan 1/2, as a result of my reconstruction of the work behind ff. 114v and 81r, by mathematical analysis of the data recorded on them. Before entering into those matters, some preliminary remarks may be useful.

For Galileo, as for Aristotle, physics was the science of nature,

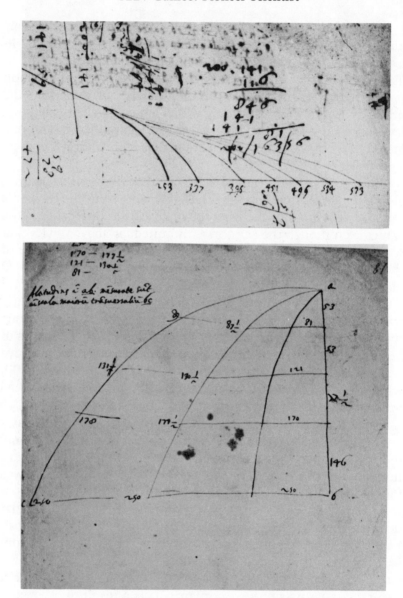

Figure 13. ff. 114v and 81r, recording experimental measurements of oblique projections.

and above all the science of natural motions – for it was Aristotle himself who wrote that ignorance of motion entails ignorance of nature. But Aristotle and Galileo were poles apart on the concept of *natural* motion. I have said several times that motion undertaken spontaneously by a body simply released from restraint was called 'natural motion,' and certainly both Galileo and Aristotle would agree to that. But Aristotle put geometrical restrictions on natural motion. For 'elemental' bodies, this was straight motion – straight down for heavy bodies, and straight up for light bodies (having levity instead of gravity). For the celestial bodies, natural motion was perfectly circular around the center of the universe, at which the earth happened to be centered also. It is rather curious that Aristotle, who did not like the excessively mathematical emphasis of his teacher, Plato, framed his own universe so strictly mathematically; and it may seem no less curious that Galileo, who is now commonly believed to have geometrized physics *a priori*, did not accept Aristotle's geometrical restrictions on natural motions.

For Galileo, the path of any natural motion was something to be discovered directly from nature. A cannonball on leaving the mouth of the cannon was no longer acted on by any force, in his view. From that moment on, it went wherever nature took it, and that was just as much a natural motion as was straight fall. He found its path to be parabolic, neglecting resistance of the air. To call that a natural motion seemed ridiculous to professors of natural philosophy; to them it was a forced motion, or mixed of natural and forced motions, and they argued about the location of the force, and whether there could be a mixed motion, and so on. Like Aristotle, they were concerned that nature be neat and orderly, obeying logical and grammatical precepts.[4] Galileo was concerned to understand as many natural motions as he could, and those that he understood lay at the heart of his mature physics.

4   Luca Valerio, the mathematician whose opinion Galileo requested concerning foundations for his new science of motion in 1609, wrote in 1612 to Galileo: 'I myself am pleased to philosophize freely and not by the rules of a certain philosophical grammar, or grammatical philosophy – if indeed that deserves to be called *philosophy* which is in use for the most part today ... '

Oblique projections of the kind described in the letter to Antonio de' Medici (shots from a cannon at any elevation) were reducible by symmetry (neglecting air resistance) to the case of level shots, and Galileo published tables of horizontal carry for shots at each elevation from 1° to 89°, calculations for which he began while still at Padua. But there is another kind of oblique projection that is not reducible to the ordinary parabola. The case Galileo considered on ff. 114v and 81r was the path of a ball that, having rolled to the end of an inclined plane, falls to the floor without deflection. The entire motion, from the moment a finger is lifted from a ball resting on an incline, is natural motion, no 'force' having acted on the ball in Galileo's sense.[5] But, understanding fully only the ball's motion before it left the plane, he sought a rule for the obliquely projected motion. Without analytic geometry, he lacked the mathematical tools for success, which is why he never mentioned the investigations. It is precisely because they did not succeed that they are of great interest, for they show how Galileo went about attacking a physical problem when it lay beyond his powers of solution.

On f. 114v Galileo sketched a diagram of an inclined plane, from the lower end of which there were seven curved lines to a horizontal below and a dotted vertical line from that to the end of the plane. Where each curved line meets the horizontal there is a number: 253, 337, 395, 451, 495, 534, 573. The angle of the plane with the horizontal is about 25° in his freehand sketch. In 1973, when I published and analyzed this as a set of oblique projections in *punti*, I calculated that 26° was the most probable slope used, and that 450 *punti* was the final drop. More probably the slope of plane was arctan $1/2 = 26.57°$, for not only is that angle just as easy to construct as 30°, but it also presented an advantage for Galileo's purposes. At the instant of projection, the horizontal component of motion would be just double the downward component, and what Galileo was seeking

---

5   Gravity was the name not of a force, but of a natural tendency downward, until
    the time of Newton.

Roll	Projection	f. 114v	Differences	
200	257	253		−4
400	337	337	0	
600	391	395	+4	
900	$449\frac{1}{2}$	451	$+1\frac{1}{2}$	
1200	493	495	+2	
1600	537	534		−3
2000	$571\frac{1}{2}$	573	+2	

were simple ratios among measurements that he intended to make.[6]

In 1982, having meanwhile found that the bronze ball must have been 20 *punti* in diameter (to the nearest *punto*) and that the width of groove was such as to reduce acceleration 5.75% from the rate for roll atop a flat plane, I recalculated from the data as seen above. The vertical drop remains 450 *punti*, but one more thing must be taken into consideration. The dotted vertical line showed that Galileo measured his distances of projection from the point vertically beneath the end of the groove. For horizontal projections, such as those recorded on f. 116v, the ball may be regarded as entering into free fall as soon as its center passes the vertical through the end of the plane (though it may rub a bit). Hence measurements of advances from the point on the floor vertically below the end of the plane are in agreement with the usual method of calculating expected results. But when the plane is appreciably inclined, say by 5° or more, that is not so. A ball of radius r will still have support until its center is past the vertical through the end of the plane by about $r \sin \theta$, on a plane at angle θ with the horizontal. In making the calculations below it was thus necessary to add that quantity to the length of roll along the plane, as compared with the round-number lengths in *punti* that Galileo had intended from his points of release, and

---

6 In fact my subsequent calculations showed arctan 1/2 to conform even better with Galileo's measurements than 26°, which I took originally as the nearest round number of degrees. The practical problem of constructing such a plane with precision did not occur to me at the outset.

also to add that amount to his measurements of the horizontal advances.

Galileo's measurements did not suggest any evident relation among the projections, though there was one striking coincidence; the roll of 900 *punti* had produced a projection of just half that length. This explains Galileo's selection of that length of roll along the plane for his next experimental measurements, in which he kept the roll constant while he varied the terminal drop. The clue to this was in a little tabulation he made on f. 114v to the left of his sketch and with the page turned sideways:

$$200 \quad - \quad 141$$
$$141 \quad - \quad 59$$
$$78 \quad - \quad 26$$

The connection that existed between these numbers is not evident because the corresponding projections were not recorded. I calculated those in the same way as before, and assuming that Galileo's experimental measurements were accurate within one or two *punti*, made the following tabulation in which the missing projections are supplied (with ±2 *punti* adjustment allowed).

Drop	Projection	[Galileo]		Drop	Projection	[Galileo]
200	257	256	—	141	199	200
141	199	200	—	59	101	100
78	126	128	—	26	51	50

What Galileo had tabulated, always for an initial roll of 900 *punti* along a slope of ratio 2:1 for horizontal advance to vertical decent, were a drop (200) that gave a projection whose half was given by another drop (78), the drop (141) for which 200 was the projection, and the drops (59 and 26) whose projections successively halved that of 200. Two calculations near Galileo's table show that he found the mean proportional of 126 and 59 as equaling 126 x 141/200, but whether he associated this quotient of 141/200 with $\sqrt{2}/2$ is dubious. No suggestive ratio had yet become apparent, and having varied rolls for a fixed drop and drops for a fixed roll, all at the same slope of plane, Galileo next tried the effect of varying the slope of his plane as well as the lengths of roll and the terminal drops.

Data from this last and most advanced series of measurements are recorded on f. 81r, which includes also a diagram and a single sentence, concerning a scaling of vertical drops that Galileo had used. That was difficult to understand, because the first and shortest drop, (53 *punti*) turned out to be actual, not scaled as were the others. My detailed analysis has been published, so here I shall merely summarize the salient points bearing on Galileo as an experimental physicist.[7]

The angles that the plane (or rather planes) made with the horizontal were arcsin 1/3, 1/6 and 1/12, or 19.47°, 9.59°, and 4.78°. Galileo had halved the steepness of roll successively, as he conceived that. Roll on the first and steepest plane was 300 *punti*, whence vertical descent along that plane to the point of oblique projection was 100 *punti*. That gave Galileo a basis of comparison for the speeds of projection, by his rule on f. 164.

For the second set of projections, from a plane only half as steep, Galileo used double the speed of projection by taking vertical descent while on the plane at 400 *punti*, the roll being 2400 *punti*. Probably the original plane used in 1604 for f. 107v had served for both these slopes, though only 2100 *punti* of its length had been used then. For the third slope, however, it was much too short. Moreover, to double the speed again would have required an impracticable length of plane, about 18 meters, so for the final set of projections Galileo halved that, reducing the roll to 9600 *punti* – still about nine meters long.

In *Two New Sciences* Galileo described his construction of a grooved plane 12 *braccia* long, which is equivalent to 7 meters. Butted together with his original plane, that would accommodate the roll of 9600 *punti*, and the accuracy of Galileo's recorded projections for the final slope is evidence that he did indeed construct that very long plane. My calculated projections, using the considerations discussed above, are compared with his data on f. 81r in the table below, from the steepest to least steep plane.

7 Cf. my 'Analysis of Galileo's experimental data,' *Annals of Science* 39 (1982), 389–97.

Drop	Projection	f. 81r	Projection	f. 81r	Projection	f. 81r
53	$81\frac{1}{2}$	81	165	$168\frac{1}{2}$	260	$257\frac{1}{2}$
100	122	121	249	$251\frac{1}{2}$	$382\frac{1}{2}$	$382\frac{1}{2}$
173	$169\frac{1}{2}$	170	348	$347\frac{1}{2}$	$526\frac{1}{2}$	$525\frac{1}{2}$

The above table shows the nearest half, because in this case Galileo recorded measurements to the half-*punto*. That he took such care is attested by his change of one number on f. 81r from $131\frac{1}{2}$ to 131. That is impressive because his procedure was a complex and difficult one. At each angle there was one drop to the floor and two drops to a board supported from the floor, for it would have been impracticable to lower the entire plane to new positions. The board must have been very carefully leveled at every position, since my calculations show that a tilt of even 1° could result in several *punti* difference for the point of impact. I presume that Galileo used an inked ball, leaving a distinctive spot at that point.

The manner in which Galileo chose distances of drop was such as to maximize his chance of detecting relations in small-number ratios, but in the end none had appeared. He summarized his data for oblique projections on f. 81r, but discarded the related notes and papers. The meaning of his notation about scaling is also of interest, but that was discussed in the paper cited, and here it suffices to say that he had become interested in drops that added equal, or nearly equal, increments of projection at the second slope as compared with the first. The work he did on horizontal projectiles at Padua is included in *Two New Sciences*, as are also the problems (as distinguished from theorems) on motion solved in 1609. What has been said above should suffice to establish Galileo as an experimental and also as a mathematical physicist before a famous event turned his attention from physics to astronomy for many years.

There is, however, one other matter that deserves mention before turning to the astronomical discoveries. Toward the end of 1608 it occurred to Galileo to link the law of fall with his discovery in 1601 that planetary distances from the sun could be related directly to orbital speeds in the Copernican system. One entry added to the top of an older diagram of concentric circles, representing planets with the sun at the center, cannot have been made before the writing of f. 91, on which Galileo had found that times in fall from rest are measures of speeds acquired. The close similarity between the form of his discovery in 1601 and that of the law of fall had led Galileo to the 'Platonic cosmogony' that he set forth near the beginning of his 1632 *Dialogue*. There he speculated that all planets were created at a place beyond Saturn (then the outermost planet known), and moved toward the sun with uniformly accelerated motion until each had reached the speed at which God had ordained it to circle the sun forever. Planetary deflection from straight accelerated to uniform circular motion was by special act of the divine will.

This speculation is wrong, but not randomly so; it is wrong by a factor of 2, so to speak. Newton, asked about it, remarked that for the sun to hold the planets in orbit in such a case, its gravitational power would have to be doubled at the instant of deflection into circular motion. Galileo certainly had no idea whatever of universal gravitation, and in the *Dialogue* he denied that he, or anyone, knew what held planets in orbit. Yet it is interesting that his mathematics of ratio and proportionality, applied to physical concepts such as distance, speed, and time, led him to a speculation that is not entirely unrelated to facts undiscovered until long after his death.

CHAPTER NINE

Astronomy and
the Telescope

Galileo's work on motion was interrupted for a time in 1604–5 by the appearance of a supernova in October, just as he was writing out for Palo Sarpi the first attempted derivation of the law of fall. Because the events bear on Galileo's Copernicanism, and are not at all widely known, it will be appropriate to review them before taking up the story of his telescopic discoveries.

A conjunction of Jupiter and Mars was predicted for 8 October 1604 by tables then in use by astronomers, and many throughout Europe were observing the skies that night. The conjunction did not take place until the following night, when a new star was seen near the two planets, approaching Venus in brightness. It was first observed at Padua on 10 October by Simon Mayr and his pupil Baldessar Capra, who confirmed its existence on 16 October after several cloudy nights. Galileo was informed of it by a friend shortly afterward.

Careful observations at Padua compared with those made at Verona soon established that no parallax was detectable. Galileo delivered three public lectures at the university, the unusual event having excited widespread interest. Cremonini, the ranking professor of natural philosophy, vigorously opposed Galileo's explanation of the implications of absence of parallax. Those had put the new star at least as far from the earth as the outer planets, if not among the fixed stars. But Aristotle held that no change could ever occur in the celestial region beyond the lunar orb, the opinion championed by Cremonini.

After a month the new star became too close to the sun to be observed, but it had already diminished in brightness. When it became a morning star about Christmas, Galileo saw that the loss of brightness (and apparent size) had continued, and he began to entertain the hypothesis that this resulted from motion away from the earth. If so, the continued lack of parallax implied motion straight away from the observer, with the possibility that some confirmation of the Copernican annual motion of the earth would result. If diminution of apparent size continued for six months, while the earth moved across the whole diameter of its orbit, a parallactic shift seemed certain to take place, measurement of which might reveal whether the earth were at rest with respect to the fixed stars or revolved around the sun.

In January 1605 a booklet was published at Padua by Antonio Lorenzini, reviewing the arguments between the 'philosophers and the mathematicians,' by which Lorenzini meant between Cremonini and Galileo; the author sided with the philosophers. One chapter was written by Cremonini, who spelled 'parallax' correctly while Lorenzini called it 'paralapse.' Galileo could not have failed to suspect Cremonini, who had consulted him about the term.[1] He composed a burlesque rustic dialogue in Paduan dialect, which he published pseudonymously in ridicule of Lorenzini's book. In it, a peasant was made to reason better than Cremonini about the new star. Two postils alluded favorably to Copernicans when this was printed at Padua in February. A second edition published at Verona a few months later changed both marginal references to sarcasms directed at the Copernicans.

This change was made because no parallactic effect at all had been observed. Commenting on this a decade ago,[2] I was then unaware that Galileo had been a semi-Copernican previously and had preferred that system until physical evidence from the tides supported annual revolution of the earth. Astronomical evidence now seemed against that, but he had semi-Copernicanism to fall

1 He had asked Galileo to explain parallax to him because he intended to write against it; Galileo was much amused that the philosopher was prepared to refute something before he even understood it.
2 *Galileo against the Philosophers* (Los Angeles 1976), 30–1. This book includes translations of two works published under pseudonyms, probably by Galileo, on the new star of 1604.

back on. It is curious that he did not simply conclude that the new star was indeed among the fixed stars and that its continued loss in apparent size did not imply motion away from terrestrial observers. The first new astronomical evidence Galileo had was not so much for the Copernican as against the Ptolemaic system. It came from the telescope, to which it is now time to turn.

Galileo did not hear of the Dutch telescope before May 1609, and probably not until July. He spoke with Sarpi at Venice about the rumors and was shown a letter from his former pupil Jacques Badovere, at Paris, confirming their truth.[3] Galileo returned to Padua and there learned that a foreigner had passed through with a telescope, going on to Venice where he hoped to sell the new device to the government. Galileo promptly started work on an instrument of his own. In his *Assayer* of 1623 he wrote of the events as he recalled them, thus:

The first night after my return I solved it [the problem] and on the following day I constructed the instrument and sent word of this to friends at Venice with whom I had discussed the matter the previous day ...

My reasoning was this. The device needs either a single glass or more than one. It cannot consist of one glass alone, because the shape of this would have to be convex ... or concave ... or bounded by parallel surfaces. But the last-named does not alter visible objects in any way, either by enlarging or reducing them; and the convex, though it does enlarge them, shows them indistinctly and confusedly. Passing then to two, and knowing that a glass with parallel faces alters nothing, I concluded that the effect would still not be achieved by combining such a glass with either of the other two. Hence I was restricted to discovering what could be done by a combination of the convex and the concave.[4]

---

3 Sarpi had learned of the Dutch telescope from a newsletter printed at The Hague in October 1608, and wrote to several of his foreign correspondents in November to ascertain the facts. By the time Galileo heard the rumors, 3x instruments made with spectacle lenses were being sold in the Paris streets, and Thomas Harriot in England was already observing the moon with his 'perspective trunk,' probably of about 8x.

4 This account, written long after the events, has been generally dismissed as fanciful at best and as not intended seriously, but it may well tell how Galileo, pressed for time as he was, came to try a convex and a concave lens in combination. The

The letter to Venice was doubtless sent to Sarpi to tell him that Galileo had confirmed the rumors: that a foreigner was about to offer an instrument to the Doge, and that he was confident of his own ability soon to make a more powerful instrument. Sarpi, asked officially for his opinion, recommended against the purchase. Galileo was indeed able to beat the competition. First he tried two spectacle lenses in a lead tube, giving about 3x, and at once began grinding a deeper concave eyepiece than used in spectacles for myopia. In the account cited, the first sentence above was followed by this:

Immediately afterward I applied myself to construction of another and better one, which several days[5] later I took to Venice, where it was seen with great admiration ... Finally ... I presented it to the Doge at a meeting of the Council.

The telescope taken to Venice late in August was about 3 feet long and magnified 8x or better. With it, Galileo sighted and described approaching ships two hours before they were seen by trained observers. Its value to Venice as a naval power was evident; in reward, Galileo's salary was raised to 1,000 florins and he was given life tenure. But there were misunderstandings, and when the award was written it obliged him to remain for life, prohibited any further increase in salary, and postponed the promised stipend until the end of his existing contract. Having received no benefit yet, Galileo felt free to negotiate for the post of court mathematican at Florence. Resentment at Venice was great when he left in September 1610.

By late November 1609 Galileo had a telescope of about 15x that he turned to the moon nightly, weather permitting, through the month of December. Early in January 1610 he had made a very good instrument of about 20x. Naming this his 'discoverer' when he later presented it to the Grand Duke of Tuscany, Galileo

Keplerian system of two convex lenses was proposed in 1611 as the result of sober optical analysis.

5 Literally, Galileo wrote 'six days,' a phrase he often used for 'several days'; 'four' similarly meant 'a few.' The chronology here can be reconstructed accurately from other documents.

also asked that it be preserved as it was, though of cardboard and not covered with decorated leather as was a telescope already left with the duke on a visit to Florence. On 7 January 1610 Galileo first sighted satellites of Jupiter with his 'discoverer' – the most important event reported in his *Sidereus Nuncius* as printed at Venice on 12 March 1610.

A few days before that famous book appeared, Galileo showed Sarpi an instrument that gave 30x magnification. That power was mentioned in the book, but I have found no evidence that Galileo used it in astronomical observation. Even in 1612, when he made his telescope an instrument of unprecented accuracy for angular measurement, he used telescopes of 18x and 20x, probably to have as great power as possible without excessively reducing the field of view. In the 1610 book he estimated the separations of satellites from one another, and from the limb of Jupiter, in 'minutes' and 'seconds,' by which he meant telescopic diameters of Jupiter and sexagesimal divisions thereof.[6]

A rather striking fact about the *Sidereus Nuncius* is that although the announcement of new astronomical discoveries provided a splendid pretext for its author to make public his Copernican preference, Galileo did not do that. He did remark that the four satellites circling Jupiter answered a common objection against the Copernican system, and in one allusion to his forthcoming *De systemate mundi* he promised arguments showing the earth to have motion. Yet apart from a single word, everything in his book fitted as well with the semi-Copernican as with the Copernican arrangement.[7] It is for that reason that it seemed appropriate above to mention the new star of 1604 and the probability that Galileo still hesitated to commit himself to anything more than the semi-Copernican system until and unless he found further and more direct astronomical evidence for annual motion of the earth.

It is also worth noting that in the printed attacks against the

6   Late in 1611 Galileo put that diameter at 50 arc-seconds, and early in 1612 measured it as 41", decreasing to 37" five months later. His procedures will be described in the next chapter.

7   Promising reasons favoring motion of the earth, Galileo used the Latin word for 'wandering,' associated especially with planets.

claimed new discoveries, Copernicanism was hardly brought up, and when mentioned at all it was never made the main issue. The published attacks were principally against the reality of Jovian satellites.[8] The most widespread oral debates arose concerning mountains and craters on the moon – long before a book on that subject was published by a Roman professor of philosophy in 1612. Perfect spherical shape of all celestial bodies was an absolute Aristotelian doctrine, even more jealously defended than the idea that all heavenly motions must be circular and have the earth as their sole center.

Galileo did not reply in print to any of the attacks, though he was ably defended in a book written by his former student John Wedderburn against the first and most vituperative attack. From Kepler, Galileo soon had strong endorsement and encouragement in an open letter, published at Prague and then reprinted (without permission) at Florence in 1610. Later the same year (though the title page shows 1611) Kepler published his corroboration of the satellites by observations of his own, using a telescope made by Galileo and lent to Kepler by the Elector of Cologne. In this booklet Kepler remarked that Galileo had given the period of the outermost satellite as 'fortnightly' (in fact Galileo had said 'semi-monthly' in his book), but that the periods of the other satellites might never be known.

Before leaving Padua for Florence Galileo made one further astronomical discovery – the strange appearance of Saturn. His telescope did not resolve Saturn's rings as such, which Galileo described as two 'ears' that he supposed to be stationary objects accompanying the planet. In July he sent a letter with a diagram to the Tuscan secretary of state, describing his discovery and requesting that this be kept secret until he published it. To the ambassador at Prague Galileo sent the jumbled letters of a Latin sentence describing the phenomenon. Kepler, who loved puzzles, did his best to discover its meaning, but in vain. The telescope that Kepler had borrowed showed planets as oblong, and the unusual shape of Saturn had not been detectable by him.

---

8 Information about the printed attacks against *The Starry Messenger* will be found in my *Telescopes, Tides, and Tactics* (Chicago 1983).

During this period Galileo continued to record positions of the Jovian satellites when weather, separation of Jupiter from the sun, and the press of other business permitted. There is no sign of his having attempted any determination of the periods of satellites before December 1610, at Florence. Venus had been too near the sun to be observed early in 1610, as it still was in July when Galileo reported the 'companions' of Saturn, saying in his letter to Prague that he had found nothing new around the other planets.[9] Venus did not appear in the evening sky before October, when Galileo began observing it, and for a month or more that planet appeared quite round to him.

During all this time the only other person in Italy known to have been making telescopes powerful enough to show Jupiter's satellites was Antonio Santini, at Venice. Santini was a retired merchant of very considerable ability in mathematics. A friend of Sagredo's, he was probably acquainted with Galileo, though no correspondence between them is known before Galileo's move to Florence. Jesuits at the Collegio Romano had been unable to see the satellites, and the professor of astronomy there, Christopher Clavius, had denied them any existence outside the lenses of Galileo's telescope. It was a common belief that all his claimed satellites were no more than optical illusions; yet for Clavius to endorse that opinion disturbed Galileo, by reason of the high respect among astronomers enjoyed by the Jesuit mathematician.

Benedetto Castelli, Galileo's ablest former student, was a Benedictine abbot then stationed at Brescia, his birthplace, who wished to come to Florence where he could work with Galileo. He hoped to have permission in due course, and meanwhile he sent to Galileo the name of a coreligionist at Milan who would receive and forward letters between them, there being no direct postal service connecting the two cities. On 5 December 1610 he wrote to Galileo saying that he had no telescope, but wondered whether Galileo had detected changes of shape in Venus such as would be expected in the Copernican system, which Castelli believed to

9 Kepler later published that letter in his own Latin translation of Galileo's Italian, mistaking the word 'around' as 'concerning' when in context it certainly had meant no other companions of the planets except for Jupiter and Saturn.

be true. That letter aroused great public interest not long ago, when a professor of the history of science declared Galileo to have cheated Castelli of an important discovery. At the same time the same charge was made in a German biography of Galileo, and it had been made previously in Germany, so it now requires refutation.

In the first place, it is hardly possible that a letter from Brescia sent to Milan for forwarding to Florence would have been in Galileo's hands in five days, as supposed by these critics. In the second place, it is not regarded as defrauding anyone to be asked whether something has been observed and to reply that it has been. Castelli's inquiry could occur quite naturally to any Copernican. What counts as a discovery is the observation, not the inquiry. In this case the verification was almost, though not quite, complete when Castelli was still writing his inquiry.

By mid-November Galileo had noticed that Venus no longer looked round, but somewhat gibbous. That proved exactly nothing, because in the Ptolemaic system Venus (and Mercury) would always appear a bit gibbous. The crucial question was whether Venus has also a crescent phase, something that was just about to be seen. Venus did not become distinctly crescent until nearly the end of December, and not until then did Galileo announce in plain words his discovery that Venus has phases like the moon's.

Now, on 11 December Galileo placed a very famous anagram at the head of a letter he sent to the Tuscan ambassador at Prague, asking that it be passed on to Kepler:

The letters transposed are these:

Haec immatura a me iam frustra leguntur

o.y.

Most illustrious Sir and high Patron

I await desirously the reply to two [letters] of mine lately written to your Excellency, to hear what Signor Kepler shall have said about the oddity of Saturn. Meanwhile I send you the cipher of another particular newly observed by me, which draws in train the decision of great controversies in astronomy, and in particular contains in itself a strong argument for the Pythagorean and Copernican arrangement; and in its time I shall publish the decipherment and other particulars.

I hope I have found the method to determine the periods of the four Medicean planets [i.e., Jovian satellites], deemed with good reason almost insoluble by Signor Kepler, to whom may it please your Excellency to pay my affectionate respect ... Excuse my brevity, as I do not feel well, and hold me in your grace for which I am anxious.

Galileo's anagram, rather ingeniously, said in translation: 'These [things], premature from me, are at present deceptively gathered together.' There were two things in the letter, totally unrelated, both 'prematurely' announced, because the work was not yet completed in either case. They were 'deceptively' combined, because neither thing threw light on the other, even if a reader did manage to unscramble the opening anagram. When properly rearranged, as Galileo disclosed in a letter dated 1 January 1611, this sentence reads: *Cynthiae figuras aemulatur mater amorum* – 'Shapes of the moon are imitated by the mother of loves [i.e., Venus].' This was singularly ingenious, in that a jumbled Latin message remained a Latin message concerning the contents of the letter on which it had preceded even the salutation.

That letter, dated 11 December 1610, alluded to Galileo's observations of the night of 10 December, which is why persons who charge Galileo with deception must assume Castelli's letter to have reached him within five days. Even if the post to Milan left Brescia the day that Castelli wrote, and the Milan-Florence post happened also to leave Milan later on the same day that the letter arrived there, and his coreligionist were waiting at the proper depot (all of which are rather unlikely), the letter would hardly reach Galileo's hands on 10 December. Had it done so, it would be a remarkable coincidence that on that night he made an observation, unrelated to Venus, that was of superlative interest to him, and to Kepler also. What that crucial observation was will be explained below.

A letter bearing the date 4 December from Venice, which did have direct postal service to Florence en route to Rome, reached Galileo on 10 or 11 December, and it supplied the motive for his urgent letter to Prague. It came from Santini, who had earlier sent one of his telescopes to Clavius, and it said, in part:

I now tell you that finally Father Clavius at Rome writes me that they have observed Jupiter, and I shall put below for you the observations copied exactly from his letter. Bit by bit the gentry are getting clear.

At the end of this letter were diagrams of the satellite positions for four nights ending 27 November. News that Jesuit astronomers at Rome had an adequate telescope could not have come at a worse time for Galileo. If they had had one earlier, he would have had their support against those who denied the reality of the satellites; a bit later, it would not have imperiled his priority in discovery of the phases of Venus. That was now the brightest object in the evening sky, and it had just reached 'last quarter,' so the next few days would be decisive. Hence Galileo's 'immature' disclosures to Prague on 11 December 1610.

What Galileo had observed on the previous night was a position of all four satellites virtually identical with their position one week (or more precisely, 167 hours) before. The only difference was that the outermost had nearly crossed its entire orbital diameter. On the page recording positions for December he marked these two entries with a cross at the end; neglecting those between them, they appear in his journal of observations thus:

Day 3	Hour[10] 5			2 ★	6	Ⓙ	4	2 ★	6	1 ★	4	3 ★	+
Day 10	Hour 4	2 ★	8	2 ★	6	Ⓙ	4	2 ★	6	1 ★			+

Galileo represented Jupiter by a circle and a satellite by a star, indicating its apparent magnitude by a number placed above the star. Below, between each adjacent pair of objects cited, he placed a number indicating the separation in visual diameters of Jupiter, counted from its limb at this time. (Early in 1612 he began showing elongations in Jovian radii, as will be explained later.)

---

10 Hours were counted from sunset of the day named.

Hence we can tabulate the satellite positions east and west of Jupiter at the above times as follows:

3 December, hour 5   E 6   Jupiter   W 4 W 10 W 14
10 December, hour 4   E 14   E 6   Jupiter   W 4 W 10

From this we can see why Galileo had recorded his nightly observations of satellite positions ever since publication of the *Sidereus Nuncius* without making any attempt to determine other orbital periods than for the outermost. That would have been inefficient before some further clue came to light. When Kepler had said in print that the other periods might never be known, that was not because the peerless calculator among astronomers could think of no way to go about the task. Galileo was aware of that when he wrote in his letter of 11 December that the problem was deemed by Kepler 'with good reason' almost insoluble. The period of the outermost satellite being known, a table of sines would suffice for eliminating that satellite from every recorded position, assuming its uniform circular motion around Jupiter. With IV eliminated, III would be the new 'outermost satellite,' whose period and separation from Jupiter could be determined just as those of IV had been, and so on. But the cumulative errors of elimination from the approximate and scanty data published in the *Sidereus Nuncius* made the procedure so unlikely to succeed that Kepler had dismissed it from serious consideration.

Galileo, who followed a very different procedure, deferred any action until he had some further clue from continued observations. He recorded apparent magnitudes because if even one or two of the satellites happened to be always brighter, or always fainter, than the others, its identification from one night to the next would be extremely helpful.[11] But it was positional information that gave Galileo the clue for which he had been waiting and that set him to work solving the problem of satellite periods.

---

11 In fact, III is brighter and IV is fainter than the others, and in the foregoing tabulation the sole first-magnitude satellite indicated was III. But with Galileo's telescope every satellite appeared faint when very close to Jupiter, so there was no decisive distinction for IV, and the recorded data for magnitudes were not useful to him at this stage.

During the long waiting period I doubt that Galileo paid any attention to observations showing fewer than all four satellites that were visible with his telescopes. The number of such cases among his records was not great, so probably no regular comparisons of them were made. On the night of 10 December, the obviously related entry was seen on the same page, a few lines above, and Galileo marked them both with crosses. They immediately suggested that, except for IV, each had circled Jupiter an integral number of times and was back where it had been a week before. Clearly the period of III was one week, very nearly; of II, half a week; and of the innermost satellite probably a quarter of a week.

The letter of 11 December was written chiefly to let Kepler know that Galileo hoped soon to have solved the problem that had appeared insoluble even to the dauntless calculator in Germany. But Santini's letter had just alerted him to the probability that Jesuits at Rome would soon see Venus pass into a crescent phase, so Galileo composed the anagram that would prove he had already predicted that, though he still did not have final ocular proof. He did not inform Clavius about the discovery until 30 December, when he also replied at length to Castelli's letter; and on New Year's Day of 1611 he sent to Prague the solution of his anagram.

The response of Kepler to news of the phases of Venus shows how greatly different his approach to astronomy was from that of Galileo. Kepler loved puzzles, and he had written down several possible solutions of the anagram, in which every planet *except* Venus had received mention. He candidly wrote to Galileo that he had always supposed, from the extreme brightness of Venus, that that planet was self-luminous, and hence could not exhibit any phases, whether or not the Copernican arrangement were correct. It seems to me interesting that Castelli, Galileo's former pupil, simply assumed the contrary without mentioning it. Kepler's idea shows that Galileo's caution, when he delayed making any positive statement until he had actually seen Venus crescent, had not been unduly conservative in science.

Nor was Galileo's apprehension that his discovery might be anticipated at Rome unfounded, though the urgency had not been as great as he thought. Later in January he received a letter from Christopher Grienberger, then assistant to Clavius and later his successor at the Collegio Romano, about the events at Rome. The telescope sent by Santini had arrived late in November. Venus, soon subjected to observation, exhibited some defect (in shape) that the Jesuits had ascribed to the instrument rather than the planet. Yet before Galileo's letter of 30 December arrived, they had realized that Venus, like the moon, sensibly lost light as it moved nearer to the sun. After the letter came, they began study of the planet's illuminated surface as compared with lunar phases, but found it hard to distinguish the exact boundary of light and darkness on Venus. Clearly, Galileo's priority of discovery had been protected primarily by his long head start in observations of Venus, and his greater familiarity with good telescopes and the interpretation of celestial appearances.

It was only after his discovery of the phases of Venus that Galileo began to speak out openly in favor of the Copernican system. Even then, well over a year elapsed before we have any documentary evidence that anyone seriously questioned the entire propriety of his preferring the new astronomy. That preference spread among his associates at Florence who appear to have begun calling themselves 'Galileists.' How the Copernican controversy arose and what parts were played in it, by Galileo and by various factions, will be deferred until later chapters. It had nothing to do with the work that Galileo immediately undertook early in 1611, which was to determine accurately the satellite periods and to construct tables of their motions.

A principal practical difficulty in the systematic procedure indicated above was that it started from maximum elongation of a satellite, the position in which the outermost is identifiable by its separation from any near neighbor. In that position it will appear stationary for a considerable time, most of its motion being toward or away from the observer. Hence the exact times of such positions are the hardest of all to determine. In contrast, times at which a satellite crosses the line from the eye to the center of Jupiter are those most exactly determinable. Galileo's first step in this direction came on the night of 29 December, when he was observing a lunar eclipse and also recorded three positions of the satellites. All four appeared at the first and last observations, eight hours apart, whereas only three were to be seen at midnight. The innermost satellite had crossed Jupiter from east to west, and Galileo estimated it to have arrived at perigee about an hour and a half before midnight. Soon afterward he made a rough table of perigeee and apogee times for that satellite (the innermost) during the next few weeks.[12]

By the end of February Galileo had identified crossings of Jupiter by satellites II and III, whose speeds near Jupiter he could compare with the speed of the innermost and which he knew to increase with size of orbit. On 15 March occurred what he called the 'great conjunction' of satellites. When he began his observations, half an hour after sunset, three satellites were seen; two were to the west of Jupiter, of which one had vanished three hours later. An hour after that none were visible, and not one had reappeared many hours later when Jupiter set. That gave to Galileo a set of satellite epochs for a single night, each at either perigee or apogee. Later he chose apogee for all epochs, but at first it was convenient in calculation to refer them all to a single base day.

Galileo had planned for some time to visit Rome, where Clavius had been expecting him. He set out on 24 March, taking his journal of observations with him, and on that night entered the record of an observation at Siena. Others were made at Santo Quirico, Acquapendente, Viterbo, Monterosi, and finally at Rome on 29 March. There, on 1 April, Galileo began constructing his table of satellite motions.

The visit to Rome in 1611 was described by a contemporary as a triumphal tour. Galileo was feted at the Collegio Romano for his *Sidereus Nuncius*. A banquet was held for him by the first true scientific society, the Academy of the Lincei founded at Rome in 1603, and there the new instrument received the name of 'telescope.' While still at Rome, Galileo was elected to the Academy. The new astronomical discoveries, including sunspots, were

---

12 The period he assumed was not a quarter-week but 41 hours, using his observations of 14 December and 6 January, separated by 554 hours. The fact that a more nearly correct period was suggested by the data for 3 December and 10 December shows how difficult the initial steps were in these pioneering determinations.

exhibited to many at Rome, where Galileo was granted an audience with Pope Paul v. From a later letter it is known that Galileo broached the subject of Copernicanism with Cardinal Bellarmine, and a document shows that the cardinal inquired of the inquisition at Venice whether Galileo had been involved in a matter affecting Cremonini.[13] It may also deserve mention that before Galileo was honored at the Collegio Romano, Bellarmine inquired of four astronomers and mathematicians there whether the claimed discoveries in the heavens were real. All attested to them, except that Clavius expressed doubt that mountains on the moon could be more than optical illusions.

Upon Galileo's return to Florence he was drawn into debate over a matter of physics, to be discussed after the astronomical calculations begun at Rome have been followed to their logical conclusion.

13 The uncompromisingly Aristotelian professor was often in trouble with inquisitors at Venice for refusing to identify in his books positions taken by Aristotle that had been ruled heretical.

CHAPTER TEN

The Telescope
and Copernicanism

The most compelling evidence that Galileo was ever to have for the correctness of the Copernican annual motion of the earth was his success in predicting eclipses of Jupiter's satellites. He never mentioned that evidence except in an appendix to his letters on sunspots published at Rome in 1613. Galileo did not become aware of the existence of satellite eclipses until July 1612, so it may appear anachronistic to mention it at this point. But its relevance is twofold: first, the principal defect of the pioneer tables of satellite motions begun at Rome on 1 April 1611 arose from Galileo's ignorance of satellite eclipses, and it was the process of refining those tables in 1611–12 that led him to compelling evidence that the earth circles the sun. Second, it would not be apparent how this 'Atlantic labor' (as Galileo called it) bore on his later Copernicanism, unless readers knew its eventual outcome from the start. Galileo was certainly unaware of any essential connection as he began the tables; to him, that was simply the kind of task astronomers were expected to perform.

It is true that discovery of the phases of Venus sufficed to induce Galileo (at long last) to make public his own Copernican preference. But the phases of Venus had nothing to do with annual motion of the earth. Provided only that Venus was inherently dark and received all its light from the sun, its phases were implied as much by the Tychonic or the semi-Copernican system as by the Copernican. Galileo can hardly have failed to know that; for him, the value of his Venus discovery was its contradiction of the cosmology of Aristotle and the astronomy of Ptolemy. That made it more probable that Copernicus was right, but it was not a compelling proof; something more would be needed for that.

The 'great conjunction' of satellites that Galileo used for his first set of epochs had not been literally a conjunction.[1] During the final two hours of Galileo's observations on 15 March 1611 satellites II and III had been eclipsed, well away from the limb of Jupiter but of course not visible. At Galileo's first observation that night, IV was west of Jupiter about one Jovian radius, a position at which Galileo's telescope might conceivably have detected it. At the end, IV was not yet quite that distance east of Jupiter, where it could not possibly be seen by Galileo.[2] The result was that he assigned to III an epoch considerably too late and a compensatory period too short in the first tables done at Rome. Apart from that, the tables were reasonably good.

Back at Florence again, Galileo used his tables of satellite motions to calculate all positions for the times at which he had recorded observations as far back as 10 November 1610. Results showed him not only that something was badly mistaken for III, but also that the farther back he went from 15 March 1611, the less accurate were his calculated positions for each satellite. A note among his papers, probably dating from July 1611, shows him to have perceived that a mistaken length of period resulted in an increasing departure from observed positions as the dates became farther from the assigned epochs, whereas a mistaken epoch produced a constant positional error. The tedious process of refinement is reflected in Galileo's papers by a succession of tables, and of parts of tables, during the latter half of 1611.

Galileo's satellite calculations until October 1611 were geocentric, not heliocentric, probably inadvertently rather than deliberately. The reason was that Galileo treated his apsidal line of Jupiter's satellites as if it supplied a fixed reference for their revolutions around the planet. In other words, it was the earth rather

---

1 It could not have been, by the rule later discovered by Laplace and used by him in establishing the stability of the solar system.

2 In a dark field Galileo's telescope could distinguish two satellites no more than 10″ apart, but near the bright disk of Jupiter it was hard for him to see one (and especially IV) less than one arc-minute from the center of the planet. Jupiter's visual radius is usually about 20″.

than the sun that in effect was treated as the center of Jupiter's own motion. In saying that the mistake was probably inadvertent, I mean that Galileo may have supposed that because he measured satellite positions with respect to Jupiter alone, it did not matter that Jupiter itself was moving. But we see the satellites as if they moved back and forth along a line through the center of Jupiter at right angles to our line of sight, and that is how Galileo measured their positions. If we observed from the center of jupiter's orbit, that line would not differ from a true apsidal line. But we do not; the heliocentric apsidal line is projected on the line along which Galileo plotted his measurements, at angles continually varying through $\pm 11°$ or a bit more.

After Galileo realized the necessity of this adjustment, his calculated positions for times at which he had recorded positions agreed quite well with earlier observations he had made. Oddly enough, however, he used Ptolemaic tables of 'prosthaphairesis,' as this was called, not the Copernican, which differ somewhat.[3] That was toward the end of 1611, when Galileo was much occupied in the dispute over floating and sinking in water that will be taken up in the next chapter. His satellite tables at this time were quite satisfactory, as shown by the following comparison of the periods he had determined with others three centuries later.

Satellite	Galileo, 1611	Samson, 1910
I	1 day 18.5 hours	1 day 18.48 hours
II	3 d 13.3 h	3 d 13.5 h
III	7 d 4 h	7 d 4 h
IV	16 d 18 – h	16 d 18+ h

The above periods of revolution as reached by Galileo late in 1611 were published in his next book. Meanwhile, early in 1612, he completely changed his way of stating orbital radii of the satellites, as explained below. His new procedure permits a com-

---

3 This choice may mean no more than that he had at hand Ptolemaic tables but not Copernican ones. Still, it bears out other evidence that Galileo was by no means a Copernican zealot by reason of the phases of Venus.

parison with modern values in terms of Jupiter's visual radius (which of course changes with Jupiter's distance from us):

Satellite	I	II	III	IV
Galileo	5.5R	9.0R	14.0R	24.0R
Modern	5.9r	9.4r	15.0r	26.4r

The reason for distinguishing the Galilean radius from ours is that his telescope showed Jupiter's disk with a spurious ring of light that extended the true disk by 10%.

To simplify the process of calculation Galileo devised an instrument that he called the 'jovilabe,' on the analogy of the Latin term 'astrolabe.' His papers include a jovilabe showing signs of much use, dating probably from March 1612, and three others reflecting its evolution from about mid-1611. The much-used example has five concentric circles drawn on heavy paper, of which the innermost represents Jupiter and the others the orbits of four satellites. All are carefully drawn to scale, as is the horizontal diameter, graduated in Jovian radii out to 24. The outermost circle, for satellite IV, is graduated in degrees from apogee (at the top), counterclockwise, and numbered at every tenth degree. Through a central hole there was a thread of which only a short piece survives. Vertical lines were drawn through the horizontal graduations, of which each fourth one was numbered along the diameter line.

From his tables of satellite motions, Galileo calculated the rotation from a satellite's epoch (always an apogee) to the date and time its position was wanted. Stretching the string to that angle as shown on the outermost circle, he noted its intersection with the circle for the satellite and read its elongation in radii at the point vertically beneath, along the horizontal diameter. Trigonometric calculations were circumvented, and the accuracy of Galileo's readings is impressive. The adjustment for parallax was of course included in the figures tabulated for degree of rotation. In 1617, probably for use on ships at sea, a brass jovilabe was made, incorporating some elaborations, now preserved at the Museum of History of Science at Florence.

The outermost circle on the earliest paper jovilabes was

marked not 24 (Jovian radii) but 15, for 'minutes' as Galileo used that term in his *Sidereus Nuncius*. It designated the visual diameter of Jupiter as estimated when counting the observed distance of a satellite from Jupiter's limb, a rough approximation by eye. In that book no distance exceeded 14 'minutes' (or indeed quite reached that); hence 15 on the first draft jovilabe was probably intended as an arbitrary outermost radius within which all the satellite orbits would fit. The change to 24 came after January 1612, when Galileo's unit became a physical entity, the visual radius of Jupiter as carefully measured – neither an arbitrary unit, nor an estimated figure in conventional circular measure.

Adoption of the new unit began on 31 January 1612, when two observations were recorded and Galileo wrote in his journal:

In this second observation was first used the [new] instrument for finding exactly the most easterly distance next [then] taken, though the instrument is not yet precisely prepared.

What Galileo had devised was a kind of micrometer, despite the fact that his lens system gave only a virtual image and the later filar or cross-hair micrometer could not be introduced into it. His device was very simple and surprisingly effective. Its origin was already foreshadowed in the *Sidereus Nuncius*, where Galileo explained how to find the power of a telescope. Two similar figures, circles or squares, were drawn side by side; the observer viewed the smaller through a telescope while looking at the larger with the free eye. When the two appeared of the same size, the ratio of their measured sizes was the magnifying power of the telescope. At least 20x, he said, would be necessary for a telescope confirming the discoveries related in his book.

When Galileo had refined his tables of satellite motions by including the adjustment for parallax, his only way to achieve still more accuracy was to make more precise measurements of the observed positions. For that he devised the superimposed-image type of micrometer to which he alluded in the previous note. It was described roughly by G.A. Borelli in 1666, long after the Keplerian lens system and the filar micrometer had replaced the Galilean (or Dutch) telescope for astronomical use. Borelli wrote that Galileo had used a scale or network, viewed together simultaneously with the satellites. That sounded implausible to me, but when I tried it I found it to work surprisingly well.[4] As I reconstruct Galileo's instrument, it consisted of a disk of cardboard on which a grid of lines 2 *punti* apart was drawn. A pin through the central intersection permitted its rotation. It was mounted on a rod, at the end of which was a ring that fitted snugly the telescope tube. Jupiter was optically centered on the pin; the disk was rotated until the satellite plane lay on a grid line. The ring was distanced so that Jupiter's disk just touched the two lines adjacent to the pin. Satellite positions were then counted off in Jovian radii as seen on that night, a physically real unit and much superior to any conventional measure. The use of such a device accounts for the uncanny accuracy of the angles Galileo drew in his journal at the end of 1612 and in January 1613 when he sighted Neptune (as a fixed star) near the satellite plane.[5]

At the bottom of the first column on the same page as the notation about the new instrument there is another note, which I believe to have been written very soon after that:

The diameter of Jupiter is to the radius of its orb as 1:275 when seen through the telescope; now if the telescope multiplies lines in the ratio 20 to 1, the true ratio of jupiter's diameter to the radius of its orb will be as 1 to 5500.[6]

Direct angular measurement of Jupiter's diameter had been

---

4   I had a low-power naval spyglass, against the end of which I held a sheet of graph paper while looking at books on shelves across the room. Optical superposition of the two images took no effort whatever.

5   In fact Neptune *was* a fixed star when Galileo first sighted it, for it was just turning from direct to retrograde motion. See S. Drake and C. Kowal, 'Galileo's sighting of Neptune,' *Scientific American* 243 (1980), 74–81. Galileo's micrometric device is also illustrated in that article.

6   Modern data for the mean orbital radius and diameter of Jupiter give the ratio 1:5412 and imply the angular diameter of the mean disk as 38.11″. In January 1612 Galileo had measured that as 41″, the disk he saw being 10% greater than actual by reason of a spurious ring of light created by his telescope. His ratio 1:275 was half the eye-to-grid distance in grid-units; see below.

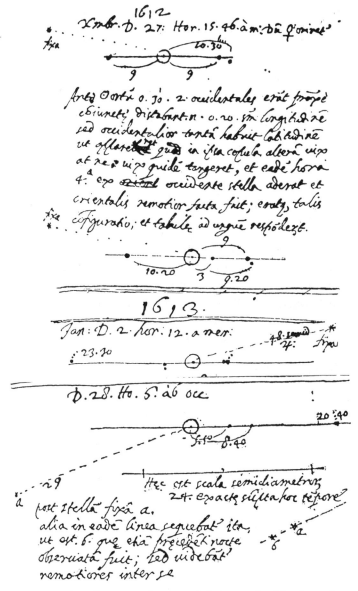

Figure 14. Journal entries for two sightings of a 'fixed star' near Jupiter in late 1612 and early 1613 that turns out to have been the planet Neptune. In the second, an actual fixed star, SAO 119234, was also recorded and designated *b*.

made possible by the superposition micrometer, for by bringing the magnified disk to fit the grid as described above (moving the grid closer to or farther from the eye), a table of sines was all that was needed. A diameter of 41″ on 31 January had dwindled to 37″ in June, when Galileo recorded a determination after Jupiter had receded from the earth. The last great naked-eye observer, Tycho Brahe, estimated jupiter's diameter at 2.5′, 225x high.

At this time Galileo was occupied not only with completing for publication his book on bodies in water, but also with his first careful observations of sunspots (which in 1611 he had regarded merely as curiosities). It was not until mid-1612 that, having further refined his tables of satellite motions and established new and more accurate epochs for each satellite, he again began calculating positions for satellites at the times of his previous observations. The first calculation, for 17 March 1612 at 6:25 PM (half an hour after sunset), agreed with his journal entry that night. The second, for 18 March at 6:27, did not; and that led to Galileo's discovery of satellite eclipses, about 15 July 1612. Briefly, the process was as follows.[7]

The original journal entry for 18 March showed observations at half an hour after sunset, when three satellites were visible, and three hours later, when but a single satellite was seen, well to the west of Jupiter. The situation rather resembled that of the 1611 'great conjunction,' so Galileo continued watching until Jupiter set, expecting to see a satellite re-emerge and perhaps enable him to refine its epoch still further. He then wrote:

I observed until the sixth hour [about midnight], nor did anything separate from conjunction with Jupiter.

This note was again before Galileo's eyes as he compared his calculation with the journal entry. The calculation disclosed that at the initial observation, half an hour after sunset, only three satellites had been distant enough from Jupiter for Galileo to have seen them, and he had recorded only three. He now saw that

7   A complete account is given in my 'Galileo and satellite prediction,' *Journal for the History of Astronomy* 10:2 (1979), 75–95.

satellite IV had been somewhat more than 2° past apogee. Had it not been for the parallax adjustment of 2° for IV in his calculation, IV would have been just 20″ beyond apogee. At once Galileo saw that it must have re-emerged far enough to be visible by the end of his long vigil, and below the calculation he wrote:

That IV was in the shadow of Jupiter clearly follows, for at the sixth hour it had not yet appeared.

Galileo would not have seen IV a half-hour after sunset and he knew that, because it was too close to Jupiter's limb by his own calculation, just emerging from behind the planet and moving eastward. But $5\frac{1}{2}$ hours later, around midnight, IV would have been well to the east, whereas all he had recorded was II, far to the west. Clearly, IV must have been receiving no light from the sun in order to have remained invisible to him. That meant that eclipses of satellites by Jupiter's interposition took place, and it also showed Galileo how he could predict them. To reflect no sunlight, a satellite must be at aphelion (to use a term coined by Kepler), a position that was easy to calculate. Galileo had only to leave out the parallax adjustment, whose meaning he now clearly saw. Omitting it was equivalent to placing the observer at the center of Jupiter's orbit; that is, on the sun; and when an observer on the sun could not see a satellite because Jupiter stood in the way, that satellite could not be receiving sunlight.

As was said at the beginning of this chapter, Galileo wrote of satellite eclipses only in an appendix to his sunspot letters. There alone did he ever assert unequivocally that motion of the earth must be taken into account in astronomy. After 1616, no Catholic was permitted to make such an assertion; hence it is no wonder that one crucial pro-Copernican argument was not included in Galileo's famous *Dialogue* of 1632. And there had been another reason for which Galileo was reluctant to say much about eclipses of Jupiter's satellites even in 1612–13.

Transoceanic navigation since the discovery of America had made it important to find some reliable means of determining the longitude of a ship at sea. Tuscany was not a leading seafaring power, but Spain was, and relations between Tuscany and Spain were of deep concern to the grand duke who employed Galileo. In September 1612 the Tuscan ambassador at Madrid presented to the Spanish government Galileo's proposal for determining longitudes at sea, which had undoubtedly been inspired by his discovery of satellite eclipses.

Longitude is just another name for time-difference between two places. The principal means of accurate determination of longitudes for very distant places had long been observation at them of lunar eclipses, using local solar times. Such eclipses do not occur as frequently as satellite eclipses, and they are less easy to time with precision. That probably started Galileo on his reflections, and he soon saw that a more generally useful, if less exact, method of establishing longitudes was more practicable. The Jovian system could be treated as a celestial clock with the satellites as its 'hands.' Galileo would prepare tables of satellite positions several months in advance, showing their times at Florence. Except for about one month each year, when Jupiter was too near the sun to be observed, those positions could be recognized by a trained observer at sea, and their times of observation would give him the longitude of the ship.

Negotiations with Spain continued for several years, Galileo offering to supply the tables and to train observers in their use, but nothing came of this. Naval officers doubted the value of the scheme, and there were real difficulties about telescopic sightings from a moving ship. Galileo offered solutions of them, but to no avail. Some twenty years later the Dutch government opened correspondence with Galileo on the matter, but he was going blind and could not make the necessary tables.[8] His tables of satellite motions and related working papers were not published until long after the work had been independently done by others, notably by Cassini after he moved to France. The first accurate map of that country was made by him, using satellite eclipses for determining longitudes of French cities. When the map was shown to Louis XIV he is said to have remarked that Cassini had deprived France of more territory than it had ever lost in wars.

8 Also, Holland was a Protestant state and Galileo refused the generous reward he was offered for technical advice because that would aggravate the inquisitors.

The appendix to Galileo's sunspot letters, printed at Rome by the Lincean Academy in 1613, contained his predictions of the satellite positions for several weeks after publication had been planned. They were impressively good, as were two of the three satellite eclipses also predicted.[9] What was printed concerning eclipses was only this:

But a more wonderful cause of the hiding of any of these is that which arises from various eclipses to which they are subject, thanks to the differing directions of the cone of shadow of Jupiter's body – which phenomenon, I confess, gave me no little trouble before its cause occurred to me. Such eclipses are sometimes of long and sometimes of short duration, and sometimes are invisible to us. These differences come about from the annual motion of the earth, from differing latitudes of Jupiter, and from the eclipsed satellite's being nearer to or farther from Jupiter, as you shall hear in more detail at the proper time.

The sunspot letters were addressed to Mark Welser, a city official of Augsburg who had sent to Galileo a little book on sunspots that he had printed on behalf of the Jesuit astronomer Christopher Scheiner. Welser asked for Galileo's opinion on it, without disclosing the author's name.[10] Scheiner believed that sunspots were dark bodies revolving between the earth and sun, saying that he was the first to discover their existence, in May 1611. Welser said he supposed they would be nothing new to Galileo, as indeed they were not. He had, however, made no serious study of sunspots until the German book made him realize the necessity of his doing so.

Galileo's long letter in reply was dated from Le Selve, the villa of Filippo Salviati about fifteen miles west of Florence, early in May. After commenting on many opinions and arguments of

9 A delay in publication thwarted Galileo's purpose, as most of the predicted events had already taken place. The eclipse that was not correctly timed, on 24 April, began when Galileo had said it would end, probably as a result of an arithmetical mistake in his calculation.

10 The book was published pseudonymously because Jesuit officials did not want the order associated with a claimed discovery that might turn out to be erroneous. For a year or more the author was called Apelles in Italy.

the author, mainly unfavourably but in a friendly tone, Galileo wrote:

Apelles comes finally to the conclusion that the spots are planets ... between the sun and Mercury or Venus. If I may give my own opinion to a friend and patron, I shall say that the solar spots are produced and dissolve upon the surface of the sun and are contiguous to it, while the sun, rotating on its axis in about one lunar month, carries them along, perhaps bringing back some that are of longer duration than a month, but so changed in shape that it is not easy for us to recognize them. This is as far as I am willing to hazard a guess at present, and I hope that your Excellency will consider the matter closed by what I have suggested.

A copy of this letter was sent a week later to Federico Cesi, head of the Lincean Academy and by this time one of Galileo's close friends and most frequent correspondents. To him Galileo wrote quite positively that the sun:

turns on itself in a lunar month with a revolution similar to those of the planets, that is, from west to east around the poles of the ecliptic – which news I think will be the funeral, or rather the extremities and Last Judgment of pseudo-philosophy ... I wait to hear spoutings of great things from the Peripetate to maintain the immutability of the heavens.

Galileo had promised Welser that he would soon send very accurate diagrams of observed sunspots for Apelles, showing their positions from day to day, elegantly drawn in a manner devised by a pupil of his. This was Benedetto Castelli, now stationed at Florence, who caught the sun's image on a circle drawn on paper and fixed at right angles to the central ray through a telescope lens. Fitting the image exactly to the circle, he rapidly drew each spot seen at an exact time. These precise records made it possible for Galileo to supply mathematical proof both of the sun's rotation and of the situation of sunspots on, or extremely close to, its surface. That was to become the heart of a second and even longer letter to Mark Welser, dated from Florence on 14 August 1612.

The existing English translation of Galileo's *History and Demonstrations concerning Sunspots* was abridged for the general reader.[11] In particular, a geometrical section was omitted from the second letter; that is supplied below, as of interest because it illustrates Galileo's approach as a scientist to the analysis of data carefully recorded and measured. Very different are the verbose arguments of the German Jesuit astronomer, who became Galileo's most vindictive enemy as a result of this book.

First your Excellency should note that the distance between the sun and us being very great in proportion to its diameter, the angle between rays from our eye to the ends of that diameter is so acute that without any sensible error we may take those rays as if they were parallel lines. Moreover, since not any two random spots are suitable for the purpose here intended, but only those that lie in the same parallel, we may choose two meeting that condition, which are known to be such that in their motion, they both pass through the center of the solar disk, or are equally distant from the same pole. Those events sometimes are found, as happens of the two spots A and B in the diagram for 1 July, where B passes close to the center on 2 July and A passes at similar proximity on 7 July, both with northerly declination. And because that distance from the center is very small, the parallel described by them is almost insensibly different from the sun's equator. So if you first imagine the line GZ representing our distance from the sun (and let Z be our eye), G is the sun's center, around which is drawn the semicircle CDE, of radius equal to (or very little less than) the radius of the circles on which I mark the spots. Hence the radius CDE will represent that which is traced by the spots A and B, which appears straight to our very distant eye at Z and lies in the same plane as the circle CLE; and the same for the diameter CGE. I say this for the reason that from the observations I have been able to make thus far, I do not know whether

---

11 In my *Discoveries and Opinions of Galileo* (New York 1957), [87]–144. The omission supplied below came at p. 109.

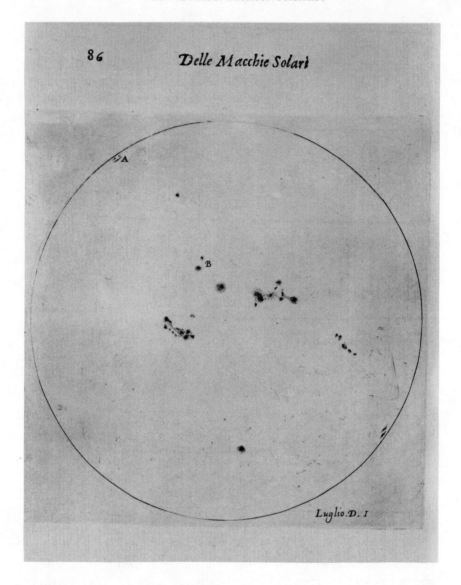

Figure 15. Sunspots on 1 July 1612, from a copy of the *Letters on Sunspots* (Rome 1613) in the University of Toronto Library.

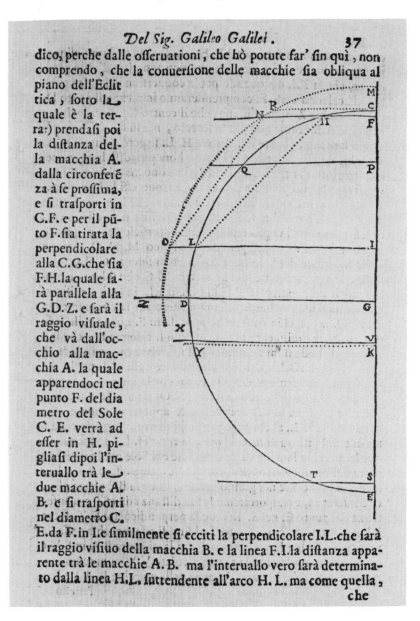

Del Sig. Galileo Galilei . **37**

dico, perche dalle offeruationi, che hò potute far' fin quì, non comprendo, che la conuerfione delle macchie fia obliqua al piano dell'Eclit tica, fotto la quale è la terra:) prendafi poi la diftanza della macchia A. dalla circonferé za à fe proffima, e fi trafporti in C.F. e per il pũto F.fia tirata la perpendicolare alla C.G.che fia F.H.la quale farà parallela alla G.D.Z. e farà il raggio vifuale, che và dall'occhio alla macchia A. la quale apparendoci nel punto F. del diametro del Sole C. E. verrà ad effer in H. pigliafi dipoi l'interuallo trà le due macchie A. B. e fi trafporti nel diametro C.

E.da F. in I.e fimilmente fi ecciti la perpendicolare I.L.che farà il raggio vifiuo della macchia B. e la linea F.I.la diftanza apparente trà le macchie A. B. ma l'interuallo vero farà determinato dalla linea H.L. futtendente all'arco H. L. ma come quella, **che**

Figure 16. Galileo's diagram for the demonstration translated here, from the above book.

rotation of the spots is oblique to the plane of the ecliptic, in which the earth lies.[12]

Now take the distance of spot *A* from the nearby circumference, carrying it to *CF*, and draw through point *F* the perpendicular to *CG*, which is *FH*, parallel to *GDZ*. That will be the visual ray from the eye to spot *A* which, appearing to us at point *F* on the sun's diameter *CE*, will be at *H*. Then take the interval between the two spots *A* and *B*, carried on the diameter *CE* from *F* to *I*, and treat similarly the perpendicular *IL*, which will be the visive ray for spot *B*, with line *FI* the apparent distance between spots *A* and *B*; but the true interval will be that determined by line *HL*, subtended by the arc *HL*. But, as that extends between rays *FH* and *IL* and is seen obliquely by reason of its slope, it does not appear different in size from *FI*.

Now when, by rotation of the sun, points *H* and *I*, dropping toward *E*, shall have point *D* in the middle, that to the eye at *Z* appears the same as the center *G*, then the two spots *A* and *B*, no longer seen [as before] in foreshortening but straight on, would appear as far apart as the subtended [chord] *HL*, if the place of the spots is on the sun's surface.

Now look at the diagram for 5 July, in which the same two spots *A* and *B* are almost equidistant from the center, and you will find their separation exactly equal to the subtended *HL*, which could in no way happen if their revolving were in a circle at all remote from the sun's surface; that will be proved as follows.

Take, for example, arc *MNO* distant from the sun's surface (that is, from circumference *CHL*) by 1/20 the diameter of its globe; extend the perpendiculars *FH* and *IL* to *N* and *O*. Clearly when the spots *A* and *B* move through circumference *MNO*, spot *A* appears at *F* when it would be at *N*, and similarly spot *B* by its appearing at *I*, would have to be at *O*; whence their true interval would be as great as the subtended line *NO*, much smaller than *HL*. Hence spots at *N* and *O*, transferred toward *E* until line *GZ* cuts the subtended *NO* in the middle perpendicularly, will have their true maximum separation and will appear much less than the subtended *HL*. Therefore the spots are not distant from the sun's surface by the twentieth part of its diameter.

---

12 This qualification is of great interest as showing Galileo to have considered the possibility around mid-1612. He did not return to investigate it until 17 years later, and then drew from the tilt of the sun's axis to the ecliptic a powerful Copernican argument, set forth in the 1632 *Dialogue*.

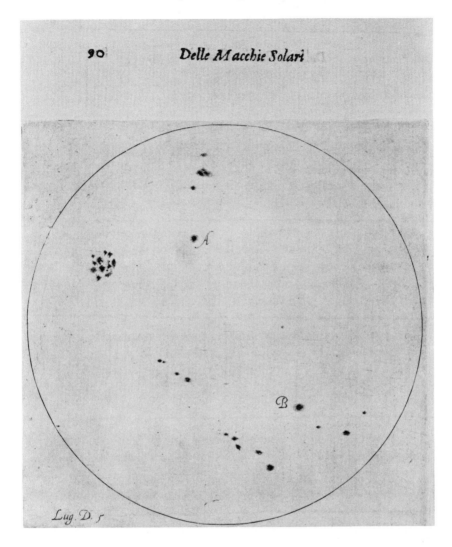

Figure 17. Sunspots recorded on 5 July 1612.

Galileo then supplied calculations for various days on which diagrams of the spots had been carefully drawn and measured, and continued:

I say further not only that the spots are very close and perhaps contiguous to the sun's surface, but in addition that they rise little above, so that their thickness, or height, is very small in comparison with their length and breadth. That is gathered from the appearance of interstices between them, often distinct right to the sun's limb, although spots are observed not far apart and along the same parallel, as happened on 26 June.

In September, before Welser received this letter, he sent to Galileo another booklet by 'Apelles' in which were included some speculations about Jupiter's satellites, Venus, and other matters to which Galileo replied vigorously on 1 December, once more from Salviati's villa Le Selve. At the end he wrote:

I shall now cease troubling your Excellency, praying you once more to offer my friendship and services to Apelles. Should you decide to show him this letter, add to it my excuses if he feels that I have dissented too violently from his views. Wishing nothing but to win a knowledge of the truth, I have frankly explained my opinion, which I am quite willing to change whenever my errors are revealed, and I shall hold myself especially obliged to anyone who favors me by exposing them and castigating me.

Before we turn to consider the dispute in physics that had taken place during the astronomical researches of 1611–12, there is one important physical statement in the second sunspot letter deserving of mention. In the context of discussing rotation of the sun, Galileo wrote:

If anyone should wish to have rotation of the spots around the sun proceed from motion that resides in the ambient and not in the sun, I think it would in any case be necessary for the ambient to communicate this movement to the solar globe as well. For I seem to have observed that physical bodies have natural inclination to some motion (as heavy bodies downward), which motion is exercised by them through an intrinsic property without need of a particular external mover, whenever they are not impeded by some obstacle. And to some other motion they have repugnance (as the same heavy bodies to motion upward), and therefore they never move in that manner unless thrown violently by an external mover. Finally, to some movements they are indifferent, as are the same heavy bodies to a horizontal motion, to which they have no inclination (since it is not toward the earth's center) nor any repugnance (since it does not carry them away from that center). And therefore, all external impediments removed, a heavy body on a spherical surface concentric with the earth will be indifferent to rest and to movement toward any part of the horizon. And it will maintain itself in that state in which it has once been placed; that is, if placed in a state of rest, it will conserve that; and if placed in movement toward the west (for example) it will maintain itself in that motion.

This conservation of motion by a body supported on a surface of spherical shape was reasserted in the later *Dialogue*, where many have mistaken it for something they call 'circular inertia' and treat as a principle that Galileo had imagined to keep a body moving in circles, whether or not it was supported. Of course he never invoked any such principle, nor had any other physicist in discussing heavy bodies near the earth.[13] Galileo had recognized for many years that natural motion of a body toward the earth's center is accompanied by increase of its speed, deducing that motion away from that center would reduce its speed; hence he concluded that only by maintaining the same distance from the center could a heavy body move with constant speed. His use of that purely kinematic proposition in the 1632 *Dialogue* has led to the nonsensical charge of 'circular intertia.' Inertia being a dynamic concept, it did not enter into Galileo's physics at all.

13 In the fourteenth century Jean Buridan remarked that planets, given an initial impetus to move around the earth, would continue forever and uniformly. But planets were then supposed to be lights embedded in transparent solid spheres, so what Buridan implied was conservation of angular momentum by the sphere, not 'circular inertia' in the motions of planets.

CHAPTER ELEVEN

Hydrostatics, Philosophers, and Religion

Galileo returned to Florence from Rome in June 1611 and was soon drawn into a controversy with professors of natural philosophy at the University of Pisa. The place was the palazzo of Filippo Salviati, a young patrician who invited Galileo to join scholars meeting occasionally at his home to discuss various topics of interest to Salviati, whose education had consisted largely of self-imposed reading. Galileo held him in the highest regard for his intelligence and learning. Salviati died in 1614 during a visit to Spain; he was immortalized as spokesman for the views of Galileo in his later books written in dialogue form.

On one occasion the topic of discussion was condensation and rarefaction of materials. Present were Professors Vicenzio di Grazia and Giorgio Coresio, both staunch Aristotelians, of whom one adduced ice as an example, saying that it was condensed water. Galileo remarked that ice would better be called rarefied water, since it floats on water. He was told that ice floats because of its flat shape, unable to overcome the resistance of water against penetration. That was the same misapprehension that had arisen in 1608 when Giovanni de' Medici and some other engineers wanted to take advantage of this imaginary resistance when designing an esplanade on the Arno River.

Galileo replied that shape was no explanation because a flat piece of ice held under water will rise through it, overcoming any alleged resistance even though its weight opposes its rising. His adversaries believed that resistance of water against motion was shown to be great by the fact that one's wrist could be hurt by striking a sword flat against water. Galileo granted that water resists *speed* of motion, though not motion as such, because even small particles will settle out from muddy water if allowed enough time. He pointed out that hardness of ice does not show it to have greater density than water, steel being harder than gold, but by no means more dense.

Matters might have ended there were it not that a Florentine who had an old grudge against Galileo was told of the debate and said he could demonstrate experimentally that shape is crucial in the matter of floating and sinking.[1] A written challenge was issued to Galileo, but the contest was not held, Galileo being warned by the grand duke not to engage in noisy public disputes, but to write out his views and arguments. The opponent intended to show that an ebony ball sinks, while an ebony lamina of the same weight can be floated on water. The original dispute, as Galileo pointed out, had to do with bodies placed *in* water, and two pieces of ebony will sink regardless of their shapes if both are literally placed in water. Though Galileo's explanation for the floating of materials denser than water in certain cases is no longer used, it deserves attention for several reasons.

Phenomena of surface tension were first scientifically analyzed in the eighteenth century; capillary action remained puzzling until the nineteenth century. Galileo's approach was necessarily very different from that of later centuries, in which the concept of force played a principal role. Nowhere is the complete absence of dynamic ideas from Galileo's mature physics more evident than in the present matter. The key concept he used was of course the principle of Archimedes, but Galileo did not simply adopt that as a principle, and he did not derive it as Archimedes had done. He based his derivation on a principle that he credited in print to Aristotle – meaning of course the author of the ancient *Problems of Mechanics*. To retrace his steps in this dispute will show how Galileo thought as a physicist in an area unrelated to

---

1 This was Ludovico delle Colombe, whose book on the new star of 1604 had been ridiculed in a book he believed to have been written by Galileo, under the pseudonym Alimberto Mauri. My reasons for believing him correct were published in *Galileo against the Philosophers*.

his new science of motion, and how that brought about, indirectly, his first brush with the Roman inquisition.

The new telescopic discoveries had disturbed astronomers primarily; natural philosophers were insulated by their treatment of
planetary systems as mathematical fictions. It was the battle over
principles of Aristotelian physics that created Galileo's first sharp
public conflict with natural philosophers. Without the resulting
open battle of books, the Copernican issue would probably not
have aroused churchmen at the time, or in the way, that the
events historically took place. Clearly hydrostatics has nothing
to do with astronomy, scientifically considered; however, history
takes it course without concern for the criteria that we ordinarily
use in our determinations of relevance. Philosophers outraged by
Galileo's hydrostatics managed to involve the Church in their
battle against his new science. Neglect of this preliminary battle-
field has resulted in a rational but unhistorical account of the
whole war, in which philosophers seem innocent bystanders.

It was precisely the Archimedean principle that seemed to be
contradicted by the floating of bodies denser than water. The
apparent contradiction had probably not even attracted Galileo's
notice until Colombe's challenge to him in the summer of 1611.
Once his attention was called to it, Galileo sought and found a
satisfactory explanation by closely studying the circumstances of
flotation by a body denser than water. When such a body floats,
it lies entirely below the surface of the surrounding water, in a
little depressed region containing not just the body, but also air.
Galileo perceived that there is no more mystery about the floating
of a chip heavier than water than there is about the floating of an
empty teakettle. Not only is the same Archimedean principle
operative; it holds in an interesting and useful way. Knowing the
specific gravity of the floating chip, the depth at which it must
float beneath the surrounding water-surface could be determined
quantitatively. Galileo was once again on home ground, because
his very first scientific work had been the making of accurate
determinations of specific gravities.

Philosophers of the period knew too little physics to think out
all the implications of Galileo's approach to the floating of bodies
in water, on which it happened that Aristotle had written com-

ments. Galileo, however, knew what Aristotle had said on the matter (in *De caelo*), having read the text with understanding. His Aristotelian adversaries had drawn a wrong conclusion from the text and adhered to it despite Galileo's best efforts to show them their error. Aristotle himself had implied exactly what Galileo had said from the very beginning: that shape affects the *speed* of motion through water, but not the *fact* of such motion. The opinions of ancient philosophers are now no longer relevant in scientific debates, but correctness about this one created the bitterness with which professors of natural philosophy fought Galileo in this matter. In their anxiety to defend an untenable position as Aristotle's own, they completely mistook the meaning of his text, and Galileo did not neglect to examine its grammar.[2]

The oral debates at Florence soon became notorious, which is why the grand duke asked his court mathematician to desist from them and instead write out his arguments. Now, when Galileo applied for the title of philosopher as well as mathematician to the Tuscan court, he had written:

As to my deserving this full title, their Highnesses may judge for themselves as soon as they give me an opportunity to deal in their presence with the men most esteemed in this subject [philosophy].

Toward the end of September such an opportunity was afforded him by the grand duke. Two cardinals happened to be at Florence and were invited to a court luncheon at which Galileo debated his science of bodies in water against the newly appointed ranking professor of natural philosophy at the University of Pisa, who had not taken part in the previous arguments. Experiments were exhibited by Galileo in support of his statements, which greatly interested members of the ruling family. One of the cardinals sided with the Aristotelian philosopher, but the other – Maffeo

2 A Roman professor of philosophy who had promised to join the fray decided not to do so, finding on closer examination that Aristotle had indeed meant the opposite of what he had first supposed. It was no wonder that those who did publish became embittered, for to admit that Galileo had understood Aristotle better than they did would be the ultimate humiliation. Being wrong about mere facts was a small thing compared with that.

Barberini, who was destined later to become Pope Urban VIII – strongly supported Galileo. Galileo had earlier written out his account of the whole affair, addressed to the grand duke, and it still survives in manuscript. After his victorious debate at the court, he put aside that manuscript and composed a book that was to have grave repercussions for his future relations with natural philosophers, who subsequently enlisted theologians on their side.

The 'discourse concerning things which stay atop water or which move in it' was written mainly at Le Selve during the first three months of 1612. Galileo took it to Florence for printing, in March, and there he found awaiting him the booklet on sunspots that Welser had sent early in January. Since Castelli had moved to Florence, Galileo asked him to see the book through the press and also to begin at once making sunspot observations from day to day, recording them with all possible accuracy. The *Discourse* is available in a new English translation with full discussion of the arguments presented in it and their background, so only a few salient points need be mentioned here.[3]

Galileo founded his whole analysis on two principles taken, he said, from the science of mechanics: that equal weights moving with equal speeds operate with equal moments, and that equal weights having unequal speeds operate with moments proportional to the speeds. Greater speed in one body can compensate for more weight in another:

This equalization between weights and speeds is found in all mechanical instruments and was considered a principle by Aristotle in his *Questions of Mechanics*. Hence we also may take it as a most true assumption that absolutely unequal weights mutually counterweigh and are rendered of equal moments whenever their weights correspond in inverse ratio to their speeds of motion.

By adopting this approach Galileo could not only derive the Archimedean principle, but also free himself from its limitations as a purely static proposition. He explained how it came about

3   S. Drake, *Cause, Experiment, and Science* (Chicago 1981)

that a very heavy beam could be floated in a few pounds of water, contrary to the old idea that nothing could lift a heavier weight than itself without the aid of some machine. Taking account of the water displaced and raised in a finite container when solids are thrust into it, Galileo appropriately modified statements by Archimedes in which volume of water was treated as indefinitely great in relation to the floating object. One form of the hydrostatic paradox was dealt with in terms of volumes of water and speeds of their rise or fall in containers having branches with differing cross-sections, implying (though not proposing) the hydraulic lift.

One of Galileo's notes implied his having a good estimate of the maximum height of the ridge surrounding a floating body more dense than water, at slightly more than an eighth of an inch. The concept of such a maximum height, beyond which water would run over the object and sink it, played a principal role in the many novel theorems proved in the *Discourse*. No less novel were several experiments that Galileo described. One such experiment was to place in water a piece of wax weighted with lead filings until it would slowly sink, but would remain afloat if gently submerged except for a part maintaining contact with the air. An inverted tumbler carrying air to it at the bottom of a tub brought the wax body up as the tumbler was raised. Demonstration of this effect at the court, and Galileo's explanation of it, caused someone who was present (probably Giovanni de' Medici) to ask Galileo if he meant to attribute a kind of magnetic power to air. Wishing not to offend him, Galileo endorsed the analogy, which led to some interesting objections and replies preserved in manuscript.

It was probably as a result of this experiment that Galileo made an interesting statement in his book: that the effort of pressing an empty tumbler into water to its brim is the same as the effort of lifting a tumblerful of water in air. He did not explain further, supposing the reason obvious. In both cases the amount of water lifted in air remains the same, and it makes no difference whether that is collected in a tumbler or is spread out over a large area. Galileo told Viviani in his old age that, in the dispute over bodies in water, ignorance had been his best teacher, for ignorance on the part of his adversaries (natural philosophers) had led him to devise experiments that he would not have bothered to carry out otherwise.

Another interesting experiment was adduced ad hominem; that is, it demolished the philosophers' assumptions without committing Galileo to any explanation of his own. Forming a wax cone that would slowly sink in water, he showed that when placed on water with the point up, it sank, but placed point down it would float, though in that position it was best fitted to pierce and divide the supposed resistance of water against motion. The *Discourse* presented theorems giving the conditions for forming such objects from materials having any assigned specific gravity.

Four books against the *Discourse* were published, one by Colombe and the other three by professors of philosophy at Pisa.[4] Galileo wrote replies on many points, to be carefully edited by Castelli, who added much more material and prepared the work for publication. But the printers at Florence appear to have been pressured not to publish the book, which was finally printed in 1615 (anonymously, though the preface was signed by Castelli). Comparable opposition had not been exerted against publication of the *Sidereus Nuncius*; indeed, no known attack was launched against Galileo's Copernican preferences before October 1612, with one possible exception.

In December 1611, after Galileo had moved to Le Selve in poor health, a friend at Rome wrote to tell him of a meeting at the home of the archbishop of Florence as reported to him by a priest who had recently returned from that city. Apparently some persons hostile to Galileo as a result of the controversy over bodies in water were present, one of whom – probably Colombe – suggested that a priest be found who would denounce Galileo from the pulpit. Certainly more than hydrostatics lay behind that proposal, which was rebuked as improper by a churchman.[5]

---

4   These men were the principals in a league formed to combat the views of Galileo. He and his friends dubbed it the 'pigeon league' because it was led by Colombe; *colombo* is Italian for 'dove.'

5   In my opinion it was the archbishop himself who did that, for it was not the priest who carried the story to Rome. The archbishop, who took a very intelligent interest in the hydrostatics dispute, was listed by Viviani among the former pupils of Galileo.

Colombe had sent to Galileo a manuscript treatise against the Copernican astronomy, making it probable that he was the person who tried prematurely to enlist a priest to denounce Galileo on that issue.

Near the end of October 1612 an elderly priest at Florence, in conversation at his convent, said that 'this Ipernicus' seemed to go against the Bible. Galileo, again at Le Selve, was told of this and wrote to learn more. The priest replied that his remark had been casual (as is obvious from his spelling of the name of Copernicus), so there was still no implied opposition from the Church. Another year passed, and then in December 1613, at Pisa, a professor of philosophy told the ruling family in Galileo's absence that motion of the earth was contrary to the Bible. That alarmed the grand duchess Christina, who asked Castelli to comment on this as a theologian. He informed Galileo, whose reply in a long letter concerning religion and science eventuated in the first truly serious events after still another year had elapsed.

In December 1614 a young priest named Caccini, of the same Dominican convent as the priest who had opposed 'Ipernicus,' in a sermon at Florence denounced the Galileists together with other mathematicians as enemies of religion. Upon word of this reaching Rome, a superior of the same order wrote to apologize to Galileo. At Pisa the older Dominican priest told Castelli of his regret that his young coreligionist had gone so far. Castelli showed him Galileo's letter of a year before, which he copied and took back to Florence. After discussions at his convent, he sent his copy of Galileo's letter to the Roman inquisition for study.

Galileo's position was that nature could not contradict the Bible, properly understood, one being God's work and the other being God's word. Words, however, may appear to say things not literally intended. Passages in the Bible had been interpreted as metaphorical in some cases. Hence in purely physical matters the Bible should be brought in last, after all factual evidence had been carefully weighed. The inquisitors turned this letter over to a qualified theologian, who reported that it contained good Catholic doctrine, although in places it might offend pious ears. After examining Caccini and two witnesses he named, the Roman inquisition dropped the matter in mid-1615.

Meanwhile Galileo had greatly expanded his 1613 letter to Castelli, citing the advice of St Augustine that no astronomical position ever be made an article of Christian faith, lest some heretic better informed in science use any error to cast doubt on doctrines properly theological. Galileo addressed this letter to Grand Duchess Christina. In 1978 Pope John Paul II, certainly a theological authority, declared that Galileo had thereby shown himself a sounder theologian than the judges who later condemned him. At the beginning of this long letter, Galileo reviewed clearly and at length the circumstances that had surrounded his writing about the subject of religion and science. The persons responsible, he said, were academic professors of philosophy who had undertaken to discredit his new discoveries in the heavens. That something more personal had come first was made clear in Galileo's account of the matter:

Persisting in their original resolve to destroy me and everything mine by any means they can think of, these men are aware of my views in astronomy and philosophy. They know that as to the arrangement of the parts of the universe, I hold the sun to be situated motionless in the center of the revolution of the celestial orbs while the earth rotates on its axis and revolves about the sun. They know also that I support this position not only by refuting the arguments of Ptolemy and Aristotle, but by producing many counter-arguments; in particular, some which relate to physical effects whose causes can perhaps be assigned in no other way ... Possibly because they are disturbed by the known truth of other [scientific] propositions of mine which differ from those commonly held, and therefore mistrusting their defence so long as they confine themselves to the field of philosophy, these men have resolved to fabricate a shield for their fallacies out of the mantle of pretended religion and the authority of the Bible. These they apply with little judgment to the refutation of arguments that they do not understand and have not even listened to.

First they have endeavored to spread the opinion that such propositions in general are contrary to the Bible and are consequently damnable and heretical. They know that it is human nature to take up causes by which a man may oppress his neighbor, no matter how unjustly, rather than those from which a man may receive some just encouragement. Hence they have had no difficulty finding men who would preach the damnability and heresy of the new doctrine from their very pulpits, with unwonted confidence, thus doing impious and inconsiderate injury not only to that doctrine and its followers, but to all mathematics and to mathematicians in general.

Next, becoming bolder, and hoping (though in vain) that this seed which first took root in their hypocritical minds would send out branches and ascend to heaven, they began scattering rumors among the people that before long, this doctrine would be condemned by the supreme authority. They know, too, that official condemnation would not only suppress the two propositions I have mentioned, but would render damnable all other astronomical and physical statements and observations that have any necessary relation or connection with those.

No one has refuted the charges thus laid by Galileo against academic professors of philosophy in 1615, and indeed they are in conformity with all the documents known to me. Historians of science have simply ignored his charge that any involvement of the Church in the matter of Copernicanism was entirely a result of concerted action by academic professors of philosophy, as things actually happened. Quite possibly Church authorities would have intervened on their own, at some later time, though all the known evidence is against that. Responsible theologians were more prudent throughout the affair than were philosophers, though in the end they, too, acted rashly, to the benefit of no one except philosophers.

In mid-1615 a Carmelite theologian at Naples published a little book reconciling a large number of biblical passages with Copernican astronomy. He then visited Rome, prepared to debate the issue with anyone. He asked Cardinal Bellarmine to comment on his book, which the noted theologian did, sending a copy to Galileo, and counseling both men to deal with Copernicanism only hypothetically and not as physically true. That sufficed for mathematicians, he wrote, while any other course would offend theologians. Reinterpretation of the Bible in this matter would be more complex than they imagined, he said, and would never be considered until and unless positive proof of the earth's motion had been found.

The position recommended by the cardinal, who was personal theological consultant to Pope Paul V, had a long and successful history extending back even farther than Christianity. Aristotle had not been dead two centuries before Greek astronomers, notably Hipparchus around 150 BC, found that measurements over four centuries and more were irreconcilable with the cosmology of the Philosopher, based on the Eudoxian homocentric spheres with the earth at the center. Eccentric circles and/or epicyclic motions of the heavenly bodies were devised, in conflict with natural philosophy. A compromise, proposed by Geminus, was accepted all around. Astronomers would confine themselves to mathematics and measurement, leaving physics entirely to natural philosophers. They were trained in causal explanation, as mathematicians were not, and in fact astronomers were in some cases better off if they did not worry about causes, said Geminus. For nearly two millennia the systems of astronomers were taken by philosophers as mathematical fictions, not as descriptions of actual motions in the heavens. Those were truly and causally performed just as Aristotle had said they were, uniformly in circles around a single fixed center of the entire universe.

Cardinal Bellarmine, who was the most highly respected of theologians, knew no reason why the same compromise that had so long kept peace between astronomers and philosophers should not serve with equal success in the seeming conflict of Copernicanism with biblical passages. Galileo did have reason to believe that, given time, science would go in the direction in which all new evidence from the telescope and from mathematical laws of motion then pointed. Common prudence dictated that the Church take no official action at the moment, while further evidence was still coming in. Any ruling in favor of Aristotelian cosmology or of Ptolemaic astronomy, as his adversaries recommended, could result in serious embarrassment for Galileo's church later on. Sure that the worst that was likely to happen was a ruling against anything but hypothetical treatment (because theologians would not reject Bellarmine's considered opinion),

Galileo journeyed to Rome late in 1615 to press for no official church action at that time.[6]

The events that followed have been of such great consequence for science and society ever since that, though they were not literally of significance to Galileo's scientific contributions, they deserve a brief review here. The Tuscan ambassador at Rome had cautioned the grand duke that 'this is not time for Galileo to come here and argue about the moon.' Paul v was not fond of intellectuals, whose opinions led to arguments within the Church. His attitude was legalistic, in accordance with his background. Galileo had various debates while at Rome in which he demolished Aristotelian ideas but gained few supporters for his own views. In February the pope, after consulting with Bellarmine, submitted two propositions to the committee of theologians known as the Qualifiers, whose duty it was to determine the orthodoxy of any disputed doctrines. Legally, their decision would settle the matter.

The first proposition was that the sun is fixed; the second, that the earth moves. It was probably the first time that the Qualifiers had been asked to decide questions of fact, because disputed propositions generally concerned philosophical opinions. They acted with such promptness that they can hardly have given serious consideration to the problems. Both propositions, they unanimously agreed (there were eleven Qualifiers), were foolish and absurd in philosophy – meaning, of course, Aristotelian natural philosophy as purged by St Thomas Aquinas from errors in Christian theology. The first proposition was judged formally heretical, as contradicting passages in Holy Scripture. Clearly the question of possibly metaphorical language, on which Galileo had particularly relied, was not even considered.[7]

6 Years ago, in my *Discoveries and Opinions of Galileo* (New York 1957), I wrote that Galileo went to Rome to oppose Bellarmine's compromise being made official. It later puzzled me that he had accepted the 1616 edict so calmly, rather as if prepared for it, and never criticized it even in private letters. The puzzles now vanish, for me at any rate.

7 For example, the passage in which the sun is said to go forth like a bridegroom from his chamber, clearly a metaphor, was literally contradicted by the first proposition.

The pope hardly needed a panel of eleven theologians to tell him that Copernican astronomy was foolish and absurd in text-book philosophy. From a legal standpoint, however, he could now put a stop to arguments among Catholic intellectuals that were annoying him. The pope was not bound by the Qualifiers' opinion about any formal heresy; only a pope or a church council could establish anything as a heresy in the Catholic faith. At the next weekly meeting of the Holy Office, Paul v told Cardinal Bellarmine to inform Galileo of the Qualifiers' decision and admonish him not to hold or defend any longer the two Copernican propositions. If Galileo resisted, the Commissary General of the Inquisition was to order him, in the presence of a notary and witnesses, that he must not hold, defend, or teach in any way the propositions that had been disqualified, lest he be imprisoned.

At the next weekly meeting Bellarmine reported that he had informed and admonished Galileo, who had promised obedience. The pope then ordered issuance of an edict by the Congregation of the Index of Prohibited Books, regulating Copernican books for all Catholics. The only book completely banned by it was the one by the Carmelite mentioned above. Two books were suspended pending certain corrections. *De revolutionibus* and a commentary on Job by a Spanish theologian. In both cases the required changes were minor. The intent of the edict, which was probably carefully written by Bellarmine himself, was to forbid any unauthorized interpretations of the Bible, and all positive assertions that the sun was fixed and the earth moved, in books written or read by Catholics.

Nothing was officially made heretical, then or later, in the Copernican matter. Nothing was changed that seriously affected astronomy, even in the most Copernican book of all. How the 1616 situation was altered in 1633 will be explained later, when the famous *Dialogue* is discussed. At that time some other events at Rome in 1616 will also be set forth. For the present it suffices to have described a situation that inhibited Galileo from writing more books for several years.

There is no reasonable doubt that the Copernican issue was precipitated by a group of natural philosophers at Florence and Pisa who leagued together during the hydrostatic controversy in opposition to Galileo and his followers. He and his friends were aware of this league, determined to oppose Galileo on everything that bore on traditional natural philosophy. What members of the 'pigeon league' saw as at stake was not just the floating of bodies in water, but sound science and due respect for authority.

CHAPTER TWELVE

Comets,
the Church,
and Tides

Robert Southwell, who was later to become president of the Royal Society, visited Vincenzio Viviani at Florence in 1661 and recorded a number of his anecdotes about Galileo. Among them was the following:

He used to say that if he were to live 1,000 years, and still expected to understand the doctrine of meteors rightly, he would consider himself the greatest fool who ever attained that age.

By 'the doctrine of meteors' was meant material of the kind that Aristotle had written separately from cosmology and physics, in the book called *Meteorologicorum*, dealing with a variety of phenomena in the elemental region below the moon. Lightning was among those, and the saltiness of seas, and comets. Aristotle held that comets were vaporous bodies near the upper boundary of the sphere of air, ignited by the sphere of fire above. Hence comets were among the things that Galileo, when he was old and blind, told his young secretary and 'last pupil' that he never hoped to understand. Such things were numerous, for Southwell entered in his commonplace book also this anecdote of his own:

When I once offered the reason [cause] of a thing to Sig. Viviani, he told me to reply to this as Galileo had done in such a case – 'that is one of so many things that I don't understand.'

The nature of comets and the causes of many things remained

unknown to Galileo not because he had not investigated them, but because he had done his best without success. Comets had indeed been the subject of Galileo's next acknowledged book after 1616, *Il Saggiatore* or 'The Assayer,' published at Rome in 1623. He had also written most of a 'Discourse on the Comets' that was published at Florence in 1619 by his friend, assistant, and once his pupil, Mario Guiducci. Both books figured in a sustained controversy between Galileo and Orazio Grassi,[1] a professor of mathematics at the Jesuit Collegio Romano, over the nature of comets and the methods proper to scientific investigation.

When Galileo returned to Florence in 1616 he resumed work on his scheme for determining longitudes at sea, further improving his tables of satellite motions and devising a brass jovilabe to be used by persons not skilled in his new science of satellite astronomy. At this time he took a house at Bellosguardo in the hills of Florence. Perhaps it was in the moving of his files of working papers that he resumed interest in the long-neglected book on motion that he was organizing for publication in mid-1609 when the telescope diverted him to astronomy. In 1618 he asked Guiducci and another student-assistant to copy theorems he had proved at Padua, one to a page, for their easier ordering and the supplying of additional theorems in logical order for the planned book. But Galileo himself did very little work on that at this time, partly by reason of illness but mainly because three comets that appeared in rapid succession late in 1618 distracted his attention once more from mathematical physics.

Comets in those days were objects of awe and dread among the general public, as signs and portents in the heavens that could be either favorable omens or warnings of pending disaster. Each was usually greeted with a rash of pamphlets by astrologers and amateur prophets, to which the comets of 1618 were no exception. Comets had also contributed to science, especially in 1577 when Tycho Brahe demonstrated that the great comet of that

---

1 Grassi has recently acquired popular fame through the book *Galileo: Heretic* by Pietro Redondi, designed to show that not Copernicanism, but atomism, was the basis of Galileo's trial and condemnation. A single undated and unsigned document addressed to an unnamed theologian is treated as exposing the scores of signed and dated records and letters as evidentially worthless.

year must have passed right through several of the supposed crystalline spheres carrying planets in their celestial courses. Aristotle was in error when he located comets below the moon, as shown by careful measurements that revealed nothing like the parallactic implications of such a view.

Among the booklets published about the 1618 comets was one by Father Orazio Grassi, intended to dispel superstitious fears and advance the scientific study of comets inaugurated by Tycho. That booklet, printed anonymously as by one of the fathers at the Collegio Romano, opened the controversy on comets that alienated the Jesuits from Galileo after years of friendly relations.

Galileo was confined to bed by illness when the comets of 1618 appeared, and they became the subject of his conversations with Guiducci and others. Several persons wrote to ask his opinion on the comets, among them Archduke Leopold of Austria, whose friendship he particularly wished to maintain.[2] Guiducci had been elected Consul of the Florentine Academy and needed a topic for his inaugural address. When the anonymous Jesuit booklet came into Galileo's hands, he suggested that Guiducci use it as a foil for discussing the comets, and a large part of Guiducci's consular address survives in Galileo's handwriting.

Two things in Grassi's booklet were particularly criticized by Galileo. One was a serious misrepresentation of the workings of the telescope, as if the distances of celestial objects could be determined by consideration of their apparent magnifications. This error had ostensibly been derived from a statement of Galileo's in his *Sidereus Nuncius*, and he wanted to correct Grassi's imputation. The other criticism, which has been almost universally distorted in histories of science, related to Grassi's unhesitating faith in assuming that terrestrial parallax was decisive concerning the location of a comet. Galileo's point was not astronomical, but methodological, and his arguments were not demonstrative, but ad hominem in the logical sense explained

---

2 Galileo hinted to him that publication of his explanation of tides, if only as an ingenious speculation, would be greatly appreciated now that the 1616 edict prevented his publishing it as a pro-Copernican argument.

earlier. Without taking any position of his own, Galileo cast into doubt the assumption adopted by the Roman Jesuit.

Parallax, Galileo agreed, is an effect having astronomical value of the highest order – when applied to determinations of the places of real physical objects. But it would be very poor judgment to apply it, say, to a rainbow, for the rainbow changes position with the observer. One could argue by pure logic that the rainbow is infinitely distant because it lacks measurable parallax, whereas in fact it is anything but that. Rainbows are reflections of refracted light from drops of water that are not very far from the eye, not even a mile away, and though halos seen around the moon are higher up, they are physically nowhere near even the moon, let alone at infinity. Hence judgment is needed in arguing by parallax or its absence. To prove in that way that a comet is located in the remote heavens, it would first have to be shown that the comet is a physical object whose place could not change with the motion of the observer.[3] That was something the Jesuit had not done, and which would be very difficult, if indeed it was possible at all.

Now, a comet might be a reflection of sunlight from some cloud of vapors in either of two ways; vapors might rise from a place on or near the earth and reach a great height, or there might occasionally be vaporous areas in the heavens that visited regions not immeasurably distant from the earth. Galileo took no affirmative position of his own on this question, though at one place he said that no reasons prevented the latter alternative. He certainly wrote nothing at all about any cone of the earth's shadow beyond which earthly vapors rose and thus became visible as comets. Yet his adversary ascribed to him precisely such an assertion, and to this day historians allege that that wild guess constituted Galileo's theory of comets. Galileo had no theory of comets, as he said in various ways in both books, and as Viviani implied in the anecdote he told to Southwell.[4]

3 The only significant motion in this respect would of course be motion of the earth, carrying the observer, but for the purely logical purposes of the argument that was irrelevant.
4 I have translated both books, annotated them, and indexed them, without ever finding any statement of the kind attributed to Galileo by Grassi (and by every

The usual statement that in the controversy over comets Galileo was in the wrong and his Jesuit opponent was in the right is made solely because most writers now assume that Galileo *must* have had a theory of comets and try to produce one to match.

One way to show that the Jesuit was not 'in the right' is to point out that the 'problem' proposed in Grassi's book was to approximate the distance of the comet from the earth – not its distance on a certain day, but its distance at all times – and he found it to be 572,728 miles. It had moved, he said, nearly uniformly on a great circle of the sphere, at the center of which he of course assumed the earth to lie. Nowhere did he consider the comet to be orbiting the sun, though that is what Tycho had concluded when he situated the 1577 comet close to the orbit of Venus, which orbit was for Tycho heliocentric.

It is hardly any wonder that Galileo did not attempt to refute all the errors in Grassi's absurd astronomical gambit. He struck instead at one single assumption, ineptly applied to some observations reported at Rome and Antwerp. That the assumption is valid when properly applied does not put Grassi in the right against Galileo, who was not obliged to accept the assumption, adopt Tycho's system, and draw conclusions. Even if he had, that might very well have more offended Grassi and his Jesuit colleagues at the Collegio Romano. There is no way to deal with pseudo-science scientifically and not alienate its supporters. Galileo may have supposed there was some hope of getting Grassi to perceive how scientific inquiry ought to be founded, and why the problem he had set himself was illusory in the first place. In any event there was some point in enabling others to see defects in the reasoning published as scientific by a professor at the renowned Jesuit college at Rome.

Grassi, however, learned nothing from the *Discourse* printed at Florence. Instead he tore from it the mask of Guiducci's name and violently attacked Galileo as its true author. That was done in a book called *Libra astronomica*, not over Grassi's true name

historian of science I have read on this issue). That was a quarter-century ago, and I know of no author who has paid any attention, let alone who has found some statement of the kind that I missed.

but as ostensibly by a Lothario Sarsi who represented himself as a student defending his teacher's reputation. At first Galileo refused to believe that the book was by Grassi, it was so inept. But the Jesuits at Rome were very proud of it, and the pseudonym left no doubt, being an anagram formed from Grassi's own name.[5]

Eventually Galileo replied to the *Libra* (meaning 'balance,' and punning from the zodiacal sign to the idea of weighing some arguments of Galileo's) with *Il Saggiatore* by applying the delicate balance used by assayers of gold. Everything said in the *Libra* was examined in detail by Galileo in his reply, which some call his 'scientific manifesto.' Not only comets and the telescope were discussed, but the structure of matter, the nature and speed of light, the multifarious sources of sound, relations of the senses to physical phenomena, uses and interpretations of experiments, the goal of science as contrasted with philosophical speculation, and the proper procedure in causal inquiries.

The *Assayer* was printed at Rome in 1623, just when Maffeo Cardinal Barberini was elected pope, taking the name Urban VIII. The event was propitious not only for Galileo but for the Lincean Academy, which was sponsoring the publication. The title page was re-engraved to show the Barberini family arms and the book was dedicated to the new pope. He was delighted with it; word soon reached Galileo that Urban was having it read to him at meals and had expressed joy on hearing that Galileo intended to visit Rome. Several members of the Lincean Academy were appointed to posts at the Vatican, including Galileo's devoted friend Giovanni Ciampoli, who was made personal secretary to the pope in charge of papal correspondence. In contrast with Paul V, Urban VIII was himself an intellectual who would advance the interests of intellectuals, as their support everywhere was desired by him. According to a later statement of Urban's, the edict of 1616 that regulated Copernican books would never have been issued if the matter had been up to him.

5 The anagram was slightly defective: Grassi was from Savona, and the letters of Horatii Grassi Savonensis transposed would not give Lotharii Sarsii Sigensani, as that appeared on the Latin title page. Galileo's friends jested that perhaps Grassi was really from Salona, noted for its fat cattle.

Galileo journeyed again to Rome in the spring of 1624 to pay homage to the new pope, who granted him six audiences during his stay. During one of those hearings Galileo outlined to him his explanation of tides, and the pope's response was later mentioned in print by a cardinal who had been present: no theory of tides could prove motions of the earth, God having had a multitude of ways to produce the same phenomena. Upon this, the scientist is reported to have fallen silent. That is not surprising, for more than one reason. Apart from the fact that one does not dispute with a pope, Galileo would in any case agree that nothing can be logically demonstrated by affirming the consequent. Moreover, if a Catholic had a demonstrative argument for motions of the earth, after 1616 he could not legitimately state it, for that would be equivalent to asserting such motions as true, a thing forbidden by the edict.

It is hardly to be doubted that Galileo hoped during his visit to persuade Urban VIII to rescind the edict, which was by this time widely misunderstood as forbidding outright all books explaining or discussing the Copernican system. Criticism of the Church on that account among intellectuals was probably frequent, especially in Germany because of Kepler's writings. But the new pope had no intention of repealing the edict, for reasons that are easily understandable. What the edict had established was a perfectly reasonable compromise according to a long-accepted tradition, by which Catholic astronomers were forbidden to assert only something they could not prove true. Hence there existed no clear reason for withdrawing the edict, and if Urban were to do that it would surely be ascribed to his personal friendship with Galileo and nothing else.

Galileo had taken with him to Rome a compound microscope he had recently perfected, which he gave to Cardinal Zollern for presentation to the duke of Bavaria. The history of microscopes having two doubly convex lenses is somewhat obscure. Such lenses were certainly uncommon, the usual burning-glass being plano-convex. In March 1620, shortly before his death, Sagredo mentioned to Galileo in a letter that he had a glass for studying details of paintings. Probably it was a large doubly convex lens similar to a modern reading-glass. Two months later Cardinal del Monte asked Galileo to make for him a glass like one brought to Florence by a painter. By 1623 Galileo was probably making the instruments that he described in a later letter, writing:

The object is attached to the rotating disk in the base so that all its parts can be seen, as but a small part can be seen at a time. The distance between lens and object must be precise, so in looking at objects that have relief one must be able to move the glass closer or farther to see this or that part; whence the tube is made movable in its base or guide. It must be used when the air is clear and bright, and best in the sunlight ... The flea is quite horrible, the gnat and the moth very beautiful; with great admiration I have seen how flies ... can walk attached to mirrors upside down.[6]

The first printed book illustrating insects seen under the microscope known to me was published at Rome in 1625, by a member of the Lincean Academy.

Cardinal Zollern, before he left Rome, had an audience with the pope. He told Urban that the 1616 edict needed delicate handling in Germany because the Protestants were all Copernicans. The pope replied that Copernicanism had not been made a heresy, and never would be, though the new system could not be proved true. Galileo may have been present, for he reported this conversation in a letter to Prince Cesi shortly before his return to Florence.

Historians have long agreed that Galileo received permission (and perhaps even encouragement) from the pope, during this visit, to write the book that was published in 1632 and was then sternly suppressed under the same pope. The evidence for this is strong, even though the two actions appear inconsistent, if not mutually contradictory. After long puzzling over this problem, I offer a story of the events that is consistent with all the documents. It cannot be told briefly, but Galileo's 1632 *Dialogue* is

6 Before Galileo left Padua in 1610 he had applied his telescope to similar observations of insects. The awkwardness of using an instrument three or four feet long for such a purpose did not prevent his giving a similar description of insects, printed in Wedderburn's reply to Horky's attack against the *Sidereus Nuncius*.

a book of such importance in the history of early modern science that I believe the circumstances of its composition, publication, and condemnation by the Roman inquisition require plausible reconstruction as fully as possible.

Galileo and the pope both knew that criticism of the Church by reason of the 1616 edict was based on the misapprehension that it forbade Catholics to write or read Copernican books. The edict did not in fact prevent the reading even of *De revolutionibus* itself, purged only of remarks in the dedicatory letter (to Pope Paul III) about biblical interpreters, and some passages in which the earth had been called a 'star,' implying its actual motion as a planet. The problem, as Urban and Galileo saw things in 1623, was to clear away the common misapprehension about the edict.

Galileo offered to compose a book in which the rival systems of astronomy would be compared. Published by church permission, after formal Catholic censorship, and written by an astronomer whose own discoveries had enlarged knowledge, such a book should silence the detractors of Catholicism and of Italian science. Arguments for and against any explanation put forth under either hypothesis would show that the edict in no way hindered Catholics in the pursuit of science.

That was an offer that Urban could hardly refuse. Galileo returned to Florence with papal gifts and a letter from the pope to the grand duke strongly commending his court mathematician. First, Galileo wrote out a reply to objections against Copernican astronomy listed in 1616 by Francesco Ingoli, head of the church agency for promotion of the Catholic faith. All the objections were answered in detail, Galileo taking the position that not weak anti-Copernican arguments, but strong theological reasons, were best in support of the 1616 edict. Ciampoli showed this reply to the pope, who appeared pleased by it. By the end of 1624 Galileo was at work on his book, which he consistently referred to as 'my dialogue on the tides' in his personal letters during the next five or six years.

It is that book which is often called Galileo's *Dialogue on the Two World Systems* in modern editions and translations. The original title page did not bear any title beyond the single word *Dialogue*. After that there were several lines of print, identifying the author by his official positions and ending with a period. A separate new sentence described what readers would find in the book, in the middle of which is the phrase now added to the title. In the original edition, discussions were promised not of the two *systems*, but of the *reasons* for and against either one, without coming to any conclusion. The modern title dates from 1744, when permission was given to reprint Galileo's *Dialogue* with an approved preface by a Catholic theologian. The printer devised the title that has been universally adopted ever since, though it was neither the title on the manuscript that Galileo took to Rome for licensing by the censors in 1630, nor the title appearing on the original edition at Florence in 1632.

All this bibliographical detail about the title of the book may seem irrelevant here, but in what follows it will be seen why Galileo's *Dialogue* will never be fully understood by readers who assume it to have been composed as a treatment of two systems of astronomy. That is what the misleading modern title says, and it has puzzled many historians that Galileo put so little astronomy in a book bearing such a title. Others find it hard to explain why its climax and conclusion is a long section about the tides, not an astronomical subject in the ordinary sense, and certainly remote from the planetary astronomies of Ptolemy and Copernicus. Hardly less perplexing has been the question why Galileo put so much physics into the book, again assuming him to have titled it as it is now published.

None of those things need be puzzling once we have become aware that Galileo's own title on the manuscript over which he labored for five or six years was *Dialogue on the Tides*. That shifts any grounds of perplexity to two proper questions – why Galileo chose that title, and how it came to be changed before his book was printed. Both those questions can be answered from clues within the book itself and from surviving letters bearing on its treatment by the censors at Rome. But first we should consider the situation as Galileo saw it when he began writing in 1624.

Some other purpose must be served by publishing a book of arguments for and against the two chief world-systems, after the Church had ruled one system a mere fiction. Without some clear purpose, arguments would appear at best to be an empty logical exercise, and at worst a veiled attack against that ruling. Galileo's intention being neither of those, he selected explanation of the tides as the vehicle in which to set forth the arguments for and against the two rival hypotheses. Astronomical eccentrics and epicycles were no less hypothetical than Copernicus's motions of the earth. As Urban VIII had said, Galileo's explanation of the tides proved nothing, being a mere consequence of a hypothesis. So much the more did it offer a valid reason for writing the book as contributing a new scientific explanation.

Once seen in this light, no puzzle remains that Galileo included an entire section on physics, in the second 'day' of the Dialogue – the longest section of all. Galileo's mature physics of relative motion, conservation of motion, and composition of motions was essential to full understanding of his explanation of tides, whereas much less would have sufficed for answering customary anti-Copernican arguments based on motions of heavy bodies. By the same token, it is hardly surprising that Ptolemaic astronomy was not even described, let alone explained in the *Dialogue*, while Copernican astronomy was represented only by a schematic diagram or two. Technicalities of planetary astronomies are entirely irrelevant to an explanation of tidal phenomena.

In short, Galileo's tidal theory constituted the organizing theme of his *Dialogue* from the very beginning, and its climax and conclusion when it was finished. It is no wonder that for years he referred to it only as a dialogue on the tides, and put that title on the finished manuscript. Neither is it difficult to see why scholars, reading the book under its present title, usually remarked that it seemed very poorly organized for an author of Galileo's talents. Some have supposed him to have tacked on an incorrect and irrelevant tide theory at the end of a book about planetary astronomy. The book simply cannot be fully understood without knowing the history of its title.

Once the organizing theme is known, evidence for what has been said is seen within the printed *Dialogue* itself. Omitting (for the moment) the author's preface, tides are first mentioned midway into the second 'day,' and only in passing. Yet near the end of the third 'day,' Galileo's spokesman Salviati said:

And since it seems to me that in these three days the system of the universe has been discussed at great length, it is now time for us to take up that principal event from which our discussions took their rise; I mean the ebb and flow of the oceans, whose cause may be assigned very probably to the movement of the earth. But this we shall postpone until tomorrow, if that is satisfactory to you.

That speech was certainly in the original manuscript, when the title was *Dialogue on the Tides*, and was inadvertently left unchanged after the title had been shortened (for a reason to be given later). The opening speeches must have brought up the tide problem, or Salviati would not have said that three whole days of discussions had arisen from it. In the printed book, however, the opening speech, by Salviati, began:

Yesterday we resolved to meet today and discuss as clearly and in as much detail as possible the character and the efficacy of those natural reasons which up to the present have been put forth by the partisans of the Aristotelian an Ptolemaic position on the one hand, and by followers of the Copernican system on the other.

Now, that speech bears the telltale mark of censorship, for let us go back a page, to the end of the author's preface. There we read, after Galileo has introduced the three interlocutors in his *Dialogue*:

Very wisely they resolved to meet together on certain days during which, setting aside all other business, they might apply themselves more methodically to the contemplation of the wonders of God in the heavens and upon the earth. They met in the palazzo of the illustrious Sagredo; and, after the customary brief exchange of compliments, Salviati commenced as follows: ...

We have just seen how Salviati did commence – not with wonders in heaven and on earth, but with astronomical systems devised by men. He could not have forgotten overnight what had been agreed, or, if he had, the others would refresh his memory.

The conflict is a glaring one, and I do not believe that Galileo can have overlooked it. With dry humor, he worked to rule here, and deliberately left the imprint of the heavy hand of censorship for readers to see. His preface had been 'adjusted' at Rome and did not arrive at Florence until printing of the text had already begun. The Roman censor instructed the Florentine inquisitor that Galileo must not make any substantial change in the officially 'adjusted' preface. Galileo gladly obliged by leaving it unaltered, not even making it fit with new opening speeches he had been obliged to write at Florence after the pope himself had ordered that tides must not appear as title and subject of the book.

Why the pope should have ordered his final change – after the meticulous censors had spent a month or more, with Galileo, going over the *Dialogue* and making alterations so that no one could show that it violated the 1616 edict in any way – was never explained; popes do not have to explain their actions. In my opinion the reason was this: The whole value to the Church of Galileo's book lay in its being published with official permission, a condition always declared on the title page itself. The pope did not want anyone to think that by approving full hypothetical discussion of motions of the earth, the Church in any way endorsed a theory of tides that was based on assuming such motions. The simplest way to avoid that was to remove Galileo's 'on the tides' from the title page on which the Church's permission would appear. Urban did not object to the inclusion of Galileo's explanation of tides; in fact, Galileo was ordered to place at its end the pope's own comment, that God had always many ways to produce any phenomena that men can observe. Urban did object against Galileo's making tides appear as subject of the book, and consequently the opening speeches had to be rewritten at Florence after Galileo received this final order upon leaving Rome in 1630.

It is not hard to reconstruct pretty confidently the nature of Galileo's original opening speeches. Salviati, recalling the agreement of the previous day, asked for suggestions about the wonder in the heavens or on earth that should open the talks. The scene being laid at Venice (and for precisely this reason), the Venetian, Sagredo, proposed the tides. Salviati then said that their friend the Academician (as Galileo was always called in the *Dialogue*) had once explained tides on the hypothesis of the two Copernican motions of the earth. Simplicio said that it would be a waste of time to discuss the consequences of an impossibility, Aristotle having proved the earth to be absolutely motionless. It was then decided to examine those proofs, and any others that could be thought of, apart from Holy Scripture. If all physical proofs against motion of the earth turned out to be inconclusive, then it would be worthwhile to examine tides on the Copernican hypothesis, for Aristotle himself had been unable to explain this great secret of nature. Such was Galileo's original plan for the *Dialogue* discussing two rival world systems without need of any conclusion in favor of either, and yet with a useful scientific purpose, as I reconstruct Galileo's original opening from the existing documents.

After the book was printed, charges were made at Rome that Galileo had violated the 1616 edict. All the precautions of the censors had not prevented that. What happened then will be taken up after a summary sketch of the composition and contents of the *Dialogue*. In particular, the story of Galileo's preface and its Roman adjustment will throw new light on the later events, and on Galileo's own character and literary skills.

CHAPTER THIRTEEN

Concerning
Galileo's *Dialogue*

A preface addressed 'to the discreet reader' that opens the famous *Dialogue* has puzzled every modern critic, and perhaps many readers at its time. It began:

Several years ago there was published at Rome a salutary edict which, in order to obviate dangerous tendencies of our present age, imposed a seasonable silence upon the Pythagorean opinion that the earth moves. There were those who imprudently asserted that this decree had its origin not in judicious inquiry, but in passion none too well-informed. Complaints were heard that advisers who were totally unskilled at astronomical observations should not clip the wings of reflective intellects by means of rash prohibitions.

Because Galileo had gone to Rome late in 1615 precisely to urge responsible church officials not to forbid that opinion, the above words may seem to have been written tongue-in-cheek at best, and at worst to have been sarcastic to the point of hypocrisy. But that was not necessarily so, for until the edict was actually decreed on 5 March 1616 its issuance could properly be opposed. Galileo alluded here to assertions about its origin and complaints about its effects made *after* the edict became law for Catholics, which Galileo reproached. Censors at Rome, who knew about his original concern over any official action, clearly did not regard this preface as written in bad faith. Even if they had not known, Galileo took care to inform them when he went on:

Upon hearing such carping insolence, my zeal could not be contained. Being thoroughly informed about that prudent determination, I decided to appear openly in the theater of the world as a witness of the sober truth. I was at that time in Rome; I was not only received by the most eminent prelates of that Court, but had their applause, and indeed this decree was not published without some previous notice of it having been given to me.

Although that was literally true, it seems sarcastic because the notice he had received (from Cardinal Bellarmine) was that he must desist from holding or defending motion of the earth, and that an edict was to be issued applying the same to Catholics in general. But again no sarcasm need be assumed, for his purpose here was to assure readers that the author was as fully informed about the circumstances surrounding the 1616 edict as any other person alive (for the cardinal and the pope had since died).

Therefore I propose in the present work to show to foreign nations[1] that as much is understood of this matter in Italy, and particularly at Rome, as trans-Alpine diligence can ever have imagined. Collecting all the reflections that properly concern the Copernican system, I shall [thus] make it known that everything was brought before the attention of the Roman censorship, and that there proceed from this clime not only dogmas for the welfare of the soul, but ingenious discoveries for the delight of the mind as well.

This statement was essential to the agreement that existed between Galileo and Urban VIII. For the *Dialogue* to serve the interests of the Church, its author had to attest that the Roman authorities responsible for the edict were not ignorant of any relevant astronomical information, and did not intend any injury to science, but had acted only for sound theological reasons. Chief among those was the suppression of unauthorized biblical

---

1 Because the *Dialogue* was written in Italian, this phrase may seem strange, but in fact it confirms my account of the events at Rome in 1623. Urban's chief concern was with the problem of Cardinal Zollern in Germany.

interpretations, synonymous at that time with Protestantism, at least in the lexicon of Catholicism.[2]

Galileo knew it to be true that church officials had acted in 1616 not to stifle science – as the academic professors of philosophy wished them to act – but to avoid opening the door to a considerable number of scriptural reinterpretations. Neither he nor the professors had got what either wished for by the 1616 compromise edict. Galileo foresaw that any official action could result in future embarrassment for his church, and argued against it, but he also knew that the action taken was the least damaging to science of any. That is why he was able so quickly to adjust to it, and never complained against it.[3]

Granting what has been said above, it will still seem odd that Galileo should have opened his preface so abruptly with a vigorous defence of an edict that he was widely known to have advised against. In all probability he did not. The printed preface had been 'adjusted' by the chief censor at Rome and sent to Florence with instructions that Galileo make no further change in it. The manuscript preface to a book titled *Dialogue on the Tides* undoubtedly opened with a paragraph explaining why a book on tides was offered to the public. There is an explanation in the printed preface, but it was put into third place, not first. After the pope's last-minute change of title and subject for the book, the ordering was changed by the chief Roman censor. Moving that paragraph was probably his principal 'adjustment.'

Now, if we read that paragraph first, we see how it had led quite naturally to a discussion of the 1616 edict.[4] Motion of the earth was to be assumed, and the question would arise whether that was legitimate in view of the 'salutary edict.' It was, so long as

2 Literally, the phrase 'dangerous tendencies,' as translated at the beginning of Galileo's preface above, represented the word 'scandals.'

3 The worst that would happen would be that Catholic scientists would in time come to appear a bit silly, insisting something to be merely hypothetical long after other scientists accepted it as completely established.

4 After promising to show how light is thrown on the tides by assuming motions of the earth, Galileo wrote: 'It is not from failing to take account of what others have said that we have yielded to asserting that the earth is motionless ... but ... for reasons supplied by piety, religion, knowledge of divine omnipotence, and consciousness of limitations of the human mind.'

motion of the earth was treated only as a hypothesis, and that was exactly what had to be made clear to those who objected and complained against the edict on the grounds named in the preface. Indeed, the whole organization of the *Dialogue* was designed to illustrate the advancement of science by means of hypotheses as well as by reasoning deductively from metaphysical principles, in the style of Aristotelian natural philosophy.

The first of the four 'days' of discussions into which the *Dialogue* was divided is, in a way, severed from the rest of the book. A long initial section embodies a critique of Aristotle's construction of his system of the world, while the Ptolemaic system as such is not mentioned, though the name of Ptolemy does occur two or three times. The balance of the first 'day,' devoted mainly to new telescopic discoveries in the heavens, has quite a long treatment of lunar phenomena and their interpretation. The separated character of this first 'day' is noted in the opening speech by Salviati in the second 'day':

Yesterday took us into so many and such great digressions twisting away from the main thread of our principal argument that I do not know whether I shall be able to go ahead without your assistance in putting me back on the track.

It was in the second 'day' that Galileo presented his mature physics of relativity of motion, conservation of motion, and the independent composition of motions, or of tendencies to motion by the same body. The presentation depended on repeated appeals to observation and experience, rather than to principles stated and adopted without proof. In contrast, some very general statements had been made in the first 'day' without proof, as for example:

Every body constituted in a state of rest but naturally capable of motion will move when set at liberty only if it has a natural tendency toward some particular place ... Having such a tendency, it naturally follows that in its motion it will be continually accelerating. Beginning with the slowest motion, it will never acquire any degree of speed without first having passed through all the gradations of lesser speed.

To prove that natural straight motion is continuously (and uniformly) accelerated from rest would have been out of place in a book addressed primarily to laymen. The proof, promised to readers in the second 'day,' required reference not just to experience and observation, but also to precise measurements in carefully designed experiments.

Continuous change was unthinkable to Aristotelian natural philosophers, and accordingly to educated people generally. It fell into the category of 'change of change' that Aristotle excluded from discussion. And it would be impossible to assign a cause for each change of speed, for example, in mathematically continuous acceleration. Galileo was fully aware of the trouble his readers would have even in grasping, let alone accepting, the kind of motion entailed by the foregoing citation from the first 'day' of his *Dialogue*. He had experienced the same difficulties himself, from 1604 to 1608. The medieval impetus-theory concept of acceleration from rest by quantum-jumps of speed was intuitive and plausible, with its succession of tiny discontinuities. It provided a cause for each change, however numerous those might be, but its implications simply were not borne out by the most careful measurements possible.

My reason for laboring this point is that Galileo's tactics in the *Dialogue* and in *Two New Sciences* are easily understood in the light of serious scientific problems that took him a long time to solve consistently with his factual knowledge, and that presented serious difficulties of clear explanation to others. Those tactics are usually ascribed to assumed fixed ideas that Galileo had been unable to shake off, which supposedly resulted in his adhering to a defective and inconsistent physics designed only to justify an oversimplified version of Copernicanism. That view, though widely favored, is supported almost entirely on the opening forty pages or so of the *Dialogue*, along with the preconception that Galileo wrote the book in an excess of Copernican zeal, and as if he never wrote anything else on physics.

Near the end of 1629 Galileo wrote to a friend that his dialogue was finished except for the opening part, in which he wished to include things more elegant than dry science. It is precisely there that we find the rather few statements he made of a metaphysical character, and it is on the basis of those that he is supposed to have revealed his superstitious awe of perfectly circular motion throughout the universe and similar unscientific prepossessions. If so, he did not appeal to any such ideas in presenting his mature physics in the second 'day,' or astronomy in the third, or tide theory in the fourth and last 'day.' He separated his metaphysics from the rest of the book, as a thing distilled from his scientific investigations, and placed it first because it was the kind of thing most likely to attract readers of his time, not because he intended to deduce his physics from it.

Aristotle also wrote his metaphysics after his physics, as shown by the name given to it (Aristotle called it 'first philosophy' or divine philosophy). In it he examined the principles he had used in the investigations he called scientific; that is, reasoning out the causes of events in nature. During the Middle Ages things got reversed; medieval philosophers considered the principles so absolutely established in Aristotle's *Metaphysics* that it was folly to allow in physics anything not deducible directly from them. Galileo saw clearly that progress in physics was barred by blind respect for Aristotle's principles, and he put into his first 'day' the following interchange between the spokesman for Aristotelians and his own spokesman:

SIMPLICIO. Aristotle first laid the basis of his argument a priori, showing the necessity of inalterable heavens by means of natural, evident, and clear principles. He afterward supported the same a posteriori by the senses and by the traditions of the ancients.

SALVIATI. What you refer to is the method he uses in writing his doctrine, but I do not believe it to be that with which he investigated. I think it certain that he first obtained it by means of the senses, experiments, and observations to assure himself as much as possible of his conclusions. Afterward he sought means to make those demonstrable. That is what is done for the most part in the demonstrative sciences ... The certainty of a conclusion assists not a little in the discovery of its proof.

The physics set forth in the second 'day' was kinematics without any trace of dynamics. The word 'impetus' is present, but not the medieval concept of an impressed force which that originally designated; Galileo redefined impetus in the first 'day' to mean speed acquired by a heavy body. Newtonian inertia is not present; it was indeed a dynamic concept, for Newton used the phrase *vis inertiae*, an inertial force. Conservation of speed sufficed in Galileo's kinematics, which was entirely consistent with Newtonian kinematics.[5] Even in his explanation of tides, for which the second 'day' was preparation, Galileo did not invoke force, but only conserved speed of water, and its obstructed flow.

There is much of interest to physicists in the second 'day' (and the fourth), of which one example will suffice here. A popular objection against daily rotation of the earth was that bodies resting on its surface would be cast off in the direction of rotation by a speed of the order of a thousand miles an hour. Copernicus himself was responsible for introducing the objection, which he mistakenly attributed to Ptolemy. After Galileo had the law of fall and perceived that speeds in fall are as the times from rest, he applied that knowledge in a purely kinematic proof that no resting body would be flung from the earth in the direction of rotation at any speed whatever, and remarked that the same would be true of a body held to the surface of a wheel not by glue, but by some tendency to move toward the center of the wheel. He also remarked that weight of the body has no more to do with these matters than with its speeds in free fall.

Euclidean geometry sufficed for Galileo's proof, in which a body was assumed to leave a rotating sphere along the tangent, at uniform speed and in the direction of rotation. A segment was cut off, from the point of tangency, and secants were drawn from points along the tangent, equally spaced and hence representing equal times. These right triangles were those of the 'triangle of speeds' in fall from rest, placed with the time-line along the horizontal instead of the usual vertical. Since the surface of the sphere always intersected the secants, the body must always have time to reach the surface – if, indeed, the body could ever leave that along the tangent, as Galileo duly questioned.

---

5 What modern historians call 'circular inertia' and ascribe to Galileo is properly called conservation of geocentric speed. Birds do not flap their wings to keep up with the earth, but to move with respect to it, as Galileo pointed out.

It is easy to confuse the idea of a body's losing all its weight at some speed of the earth's rotation with the idea of its being flung off, and many commentators have vainly sought some flaw in Galileo's geometrical and kinematic proof. The fact that a body made weightless in this way would simply be in orbit, so that no more speed could be imparted to it by still swifter rotation, became overlooked – and along with it the fact that the orbiting body would then appear to be moving *opposite* to the direction of rotation. Of course nothing of this kind occurred to Galileo, who had at the outset dismissed the imaginary case of a completely weightless body from consideration. When I translated the *Dialogue* thirty-five years ago, I too looked for a flaw in Galileo's argument, so deeply ingrained is the notion that little can be done in physics except by dynamics, and the notion that Galileo spent more time trying to justify Copernicanism than he spent thinking as a physicist.

Related to Galileo's argument about projection is a curious speculation, earlier in the second 'day,' known as 'semicircular fall.' Although Galileo expressly considered fall from the top of a tower only to its base, his diagram included construction lines to the earth's center. Mersenne supposed Galileo to have imagined that a body falling from any height, even from the moon, would take the same time, and Einstein expressed surprise to me that Galileo seemed to have the body stop at the center of the earth, whereas by his own law of fall it would have attained very high speed. Those puzzles vanish when the text is read carefully and without ascribing to Galileo any more than he asserted, which is interesting enough by itself. It is very nearly true, he said, that a body seen by us to fall straight truly falls in a circular motion, as it also moved when resting on the tower. Moreover, it moves not a whit less when on the tower than while falling to its base. And finally, the true and real motion is not accelerated, but uniform motion[6] (for Galileo, uniform circular motion was not an accelerated motion, as it became in Newtonian dynamics).

---

6 For these delightful 'bizarries,' as Galileo called them here, or 'jocular speculations,' as he said in a later letter, he assumed equal times along a circular arc. That hardly proves Galileo to have believed in 'circular inertia,' especially as he was talking about straight fall as observed on a rotating earth and asking how that would appear to an observer not sharing in its rotation.

The second 'day' mainly concerned arguments relating to the earth's daily rotation; the third, arguments for and against its annual revolution, or relating to both rotation and revolution of the earth. Accordingly, the second 'day' was chiefly physics and the third involved astronomy. But what was discussed there was not planetary astronomy in the sense that interested Ptolemy, or Copernicus, or Tycho, or Kepler. Galileo told his readers about new stars, telescopic implications for stellar and planetary magnitudes, sunspots, and possible ways of detecting stellar parallax that had not been tried. He added a section on magnetic phenomena, and what was implied about 'earth' as an 'element.' Not astronomy as it had always been pursued, but vistas in a new astronomy from which physics would no longer be barred, unfolded in the third 'day.'

Galileo's astronomical contributions were considerable, although he did nothing notable in planetary astronomy. Satellite astronomy, which he pioneered to the extent that it is hardly too much to say he invented it, was not even mentioned in the *Dialogue*. The third 'day' opened with a crushing indictment of a book pretending to prove that Tycho's star (the supernova of 1572) was located below the moon. The author[7] had selected some dozen observations that implied a relatively large parallax for the nova. Galileo calculated for all the recorded observations, of varying reliability, the adjustments to each that would bring them into agreement on one location for the star. To get ten in agreement on placement just barely below the moon, the necessary adjustments totaled 756 arc-minutes. In contrast, five other observations put the star among the fixed stars, while five more required only a total of 10.5' adjustment to place it there. If I am not mistaken, the concept of probable error, essential to modern physical science, had a beginning in Galileo's *Dialogue*.

Also in the third 'day,' Galileo explained the telescopic appearance of planets as well-defined disks, whereas even first-magnitude fixed stars were reduced to shapeless spots of light. He described the use of a string to hide a star, taking the ratio of

7   Scipio Chiaramonte, of whom Galileo had written favorably in *The Assayer*. Kepler correctly predicted that Galileo would not long continue to approve of him.

thickness of the string to its distance from the eye for estimating the star's angular diameter, which he put at not more than 5″ of arc. The diameter of the pupil was mentioned, along with an adjustment for taking that into account.

As for stellar parallax, Galileo first analyzed just what effect had the best chance of being detectable and named Vega as the star most deserving of careful study. On a suitably chosen hill, a fixed upright post could be placed that would just occult the star. Semiannual sightings with a telescope, left fixed in the plane below, might reveal parallax arising from observations at positions thus separated by the earth's orbital diameter.

But by far the most interesting and significant section in the third 'day' concerned evidence from sunspots for motion of the earth. The argument did not occur to Galileo until early in 1629. Castelli, who had been called to Rome by Urban VIII to superintend hydraulic works there, had completed his book on the measurement of flowing waters and sent a copy to Galileo about the end of 1628. In the ensuing correspondence he mentioned that the German Jesuit Scheiner (the 'Apelles' of 1612) was completing an enormous book on sunspots, begun at Rome in 1626. In fact, the printing of it was not completed until mid-1630, after Galileo's *Dialogue* had been approved by the censors subject to the various alterations that Galileo was to make at Florence.

The news about Scheiner around the beginning of 1629 did not impress Galileo, who believed he had said everything worth saying about sunspots. But then Castelli described a particularly large sunspot he had observed until it passed from view on 9 January, and again when it reappeared on 24 January. Galileo appears to have reread his correspondence of the earlier period, and there he found a letter describing a sunspot phenomenon that he had not observed. The letter had been written from Paris by Francesco Sizzi, a young Florentine, to a Dominican father at Rome, who sent it on to Galileo as showing that Sizzi, who had published against Galileo's first telescopic discoveries, had now been won over by his books on floating bodies and on sunspots.

Sizzi and some mathematicians at Paris had observed that the shape and orientation of sunspot paths changed throughout a year. His description in the letter was a bit vague, but it was enough to induce Galileo to consider the possibility that the sun's axis of rotation was not at right angles to the ecliptic, as he had thought, but was in fact tilted from that. The geometry of that situation, even without knowledge of the angle of tilt or the times of year when our line of sight lies in the plane of the solar axis of rotation, provided Galileo with a new and powerful argument that the earth has at least one of the two Copernican motions, and probably both.[8]

Galileo's correspondence of 1629 shows that he had for some time neglected work on his *Dialogue*, suddenly resuming it with renewed energy and intending to finish it that year. For the first time he mentioned in a letter that his book on tides would contain ample confirmation of the Copernican system. Near its end he had Sagredo, spokesman for the open-minded reader, say:

In the conversations of these four days we have, then, strong evidence in favor of the Copernican system, among which three have been shown to be very convincing – those taken from the stoppings and retrograde motions of the planets and their approaches toward and recedings from the earth; second, from the revolving of the sun and what is seen in the sunspots; and third from the ebbing and flowing of the ocean tides.

It is true that the sunspot argument was a late addition to the *Dialogue*, but it is quite impossible that Galileo took his clue to it from Scheiner's recent book on sunspots, as Scheiner was quite convinced had been the case. By no means could it have escaped the censors if Galileo had added a principal argument without authorization, after provisional permission had been given for publication of the *Dialogue*. But Scheiner's groundless rage, in the opinion of well-informed persons at Rome, was responsible for inaugurating the drastic action taken there against Galileo.[9]

---

8 Galileo, *Dialogue* (Berkeley 1953), 346–55. The power of the argument is clearly seen when it is remembered that the adversaries assumed the earth to be absolutely motionless (whatever that means) and held the sun to move around it in perfectly uniform and perfectly circular motion.

9 It is probable that Scheiner, who moved to Rome about the time *The Assayer* was published and was angered by its opening pages, had a great deal to do with Grassi's reply to it, which Grassi was reluctant to publish.

The fourth 'day' was devoted entirely to Galileo's theory of the tides, much expanded beyond its original summary in Sarpi's notebooks of 1595, and further improved over Galileo's manuscript written for a friendly cardinal at Rome early in 1616. The main addition was an explanation of spring and neap tides by change of the moon's position in the earth-moon system, varying the speed of the earth in its orbit. Although some unchanging force was assumed to move the two bodies around the sun, Galileo's account of the effect on seas remained kinematic. So was his reasoning about equinoctial and solstitial tides by variations of the two components in the 'primary cause' of disturbance of large seas.[10]

Various complaints were filed with the authorities at Rome against the *Dialogue*. Most of them charged that Galileo had gone beyond the bounds of hypothetical argument, especially in the matter of explaining tides. From this it is clear that even his adversaries understood the 1616 edict as it was intended, and as the official censors had interpreted it for fifteen years. Only the positive assertion that the earth moves was forbidden, not this assumption in scientific arguments. The pope appointed a panel to examine the charges, and they reported to him that if there were any passages of that kind, they could easily be corrected if the book had value for the Church. In addition to such charges, however, they noted a personal accusation against Galileo that infuriated the pope against him for life.

It had been alleged that in 1616 Galileo was personally ordered by the head of the Roman inquisition never again to hold, defend, or 'teach in any way' the Copernican doctrine, lest he be imprisoned. In fact, a document to that effect was found in the records of the Inquisition. No one, however, had ever signed it, and because it was a notarial memorandum it would have had to be signed by everyone named in it to have any legal force. Galileo was ordered to Rome to stand trial under 'grave suspicion of

10 Galileo, depending on Strabo's description of those tides, had their sizes reversed. They are barely perceptible in the Mediterranean Sea, with whose hydrology Galileo displayed an accurate knowledge.

heresy.'[11] At the trial he produced an affidavit by Cardinal Bellarmine, dated and signed in 1616 and written in the cardinal's own hand, that completely discredits the alleged order to Galileo personally. If ever given, it was given illegally.

The impasse created by the affidavit was resolved by extrajudicial negotiation. The commissary then in office persuaded Galileo to confess to a lesser offense, admitting that he had gone too far in his book but without any intent to disobey the edict. Expecting lenient treatment in return, Galileo was shocked to be sentenced to indefinite imprisonment and to have his book burned by public executioners wherever a copy was found. The offense of which he was found guilty was second only to heresy itself in gravity under Catholic law.

By that action, signed by seven of the ten cardinals of the Inquisition who sat as Galileo's judges, Copernicanism became for the first time tantamount to heresy for Catholics generally. The verdict was what is now called judiciary legislation. The police arm of the Church had repudiated the actions of its own officers, the licensers of books. After 1633 this unjust verdict replaced the edict that had been carefully worded (probably by Cardinal Bellarmine himself) to preclude interference with the advance of science in Italy, and among Catholics elsewhere. Aristotelian natural philosophers were handed on a platter in 1633 what they had been refused by prudent church authorities in 1616.

In 1635 the *Dialogue* was translated into Latin and published at Strassburg by the famous Dutch firm of Elzevir. A few years later that was reprinted in France, and then again in England. One English translation had been promptly made, and another was published in 1661. The original Italian text was reprinted with a false imprint (of Florence) at Naples in 1710, and finally, in 1744, the Church permitted it to be printed with an explanatory theological preface. The *Dialogue* was removed from the index of prohibited books after 1818, when a daring Catholic astronomer published, after his having been refused official permission, a

11 Disobedience of a personal order given by the Holy Office was prima facie evidence of heresy.

textbook in which motions of the earth were asserted without any pretense of hypothetical treatment.

That Galileo had acted in exact compliance with the Roman authorities when he published the *Dialogue* is clear from a letter written in 1631 by the Master of the Holy Palace at Rome to the chief inquisitor at Florence. The same letter confirms various statements made above about the original title, the legitimacy of hypothetical treatment of Copernicanism until 1633, and the ways in which Galileo's book would serve the interests of the Church. This letter was written because an outbreak of plague had closed the roads to Rome; hence Galileo could not return there for final approval of the changes ordered, and if the manuscript were sent to Rome it would be subjected to fumigation page by page. What the chief censor at Rome wrote was this:

Sig. Galileo thinks of printing a work of his that formerly had the title *De fluxu et refluxu maris*, which discusses this [the tides] with probable reasons taken from the Copernican system according to mobility of the earth, and he claims to facilitate understanding of that great secret of nature by this position, corroborating it [that system] vividly with this use. He came here to Rome to show the work, which was signed by me assuming accommodations that had to be made in it, and his bringing it back to receive final approval to print. Being unable to do that because of closing of the roads and danger to the original, and the author wishing to bring this business to an end, your Reverence may use your authority and approve or not approve the book independently of my revision, but keeping it in mind that it is his Holiness's will that the title and subject may not propose the tides, but absolutely mathematical considerations of the Copernican position about motion of the earth, with the purpose of proving that, excluding divine revelation and holy doctrine, the appearances could be saved in this position, [he] resolving all the contrary arguments that might be adduced from experience and the Peripatetic philosophy so that this position is never conceded absolute, but only hypothetical, truth and apart from the Bible. It must also be shown that this work was done only to show that all the arguments which this side can adduce are known [here], and that it was not from lack of their knowledge at Rome that this opinion was abandoned, conformably with the beginning and the end of the book that will be sent from here adjusted. With the above caution the book will have no obstacle here in Rome, and your Reverence may pacify the author and serve his Highness [the Grand Duke] who shows much pressure in this matter.

Probable reasoning was never confused with demonstration or proof, as it often is now, so the common modern portrayal of the *Dialogue* as Galileo's vain attempt to *prove* the Copernican system is simply mistaken. His *Dialogue* was a sustained exercise in the weighing of all known evidence and judging where preponderance lay, apart from the ultimate authority of the Church.

CHAPTER FOURTEEN

Two New Sciences

While Galileo was at Rome awaiting trial he received an invitation from Archbishop Ascanio Piccolimini of Siena to visit him on his way back to Florence. Galileo was emotionally disturbed after hearing the sentence pronounced against him. The Tuscan ambassador applied immediately to the pope for pardon, or at the least for commutation of sentence so that Galileo would not be held in the dungeons of the Roman inquisition. The pope agreed to allow Galileo to proceed as far as Siena, to remain there in custody of the archbishop pending further instructions. It was there that he began writing *Two New Sciences*, his last and scientifically his most important book.

During the period between revision of the manuscript of his *Dialogue* and its printing, Galileo had again taken up his notes on motion, neglected for many years, with the idea of ordering them suitably for publication as a book. Without those notes at Siena, he started a related project by writing on matter rather than on motion, and in particular on strengths of material structures. When *Two New Sciences* was published, in dialogue form, its first two 'days' comprised that new science, followed by two 'days' on the new science of motion completed at Arcetri, in the hills near Florence. Galileo had been permitted to leave Siena near the end of 1633, and lived the rest of his life under house arrest at his villa in Arcetri.

Theorems on the breaking strengths of beams occupy the second 'day' of *Two New Sciences*. The first 'day' opened at the arsenal in Venice with discussion of the question why large vessels had to be supported by very massive structures while they were being constructed, though smaller boats needed little or no special support. The concept of things breaking under their own weight led on to consideration of the greater relative strength of small animals than large ones, the limiting size of a tree, or of a giant, and the reason for which sea animals may be enormously larger than any land animal. Those discussions led naturally on to speculations about the structure of matter and the nature of continuous magnitudes.

Motion also entered into the opening 'day,' in the course of which Galileo demolished the Aristotelian theory of speeds in fall. That had two parts, the first being that speeds are as the weights of falling bodies. Refutation was conducted in a way that shows why Galileo was detested by natural philosophers, who saw that the same argument should have occurred to Aristotle, or long ago to any able student of his works. Philosophers might have forgiven Galileo for pointing out a fact that contradicted Aristotle, but to reveal him as having fallen into a bald self-contradiction was insupportable. That is what Galileo did, and his argument makes it appear as if no universe could exist in which bodies would fall with speeds proportional to their weights, rather than just that we happen not to live in such a universe. The laws of physics would of course be different from ours, but no self-contradiction ought to follow from Aristotle's rule; it merely fails to agree with actual measurements here on earth. It is of interest to see how Galileo proceeded, and why that made both Aristotle and all his followers look ridiculous.

Galileo's argument can be paraphrased thus. Assume two stones, the weight of one being double that of the other. The heavier, according to Aristotle, will fall twice as fast as the lighter. Assign them speeds of 4 and 2; tie them to the ends of a stout cord and drop them. The slower-moving stone ought to drag back against the faster one, its natural speed being no more than 2, and so the net speed should be less than 4. Now shorten the cord until the stones are in contact. What is equivalent to a stone of greater weight than required for speed of 4 will then be falling with a speed less than 4. In that case Aristotle's rule entailed a contradiction, or at any rate he had neglected to explain something that any alert philosopher ought to have noted long before Galileo's time.

The other part of Aristotle's rules for speed of fall was that speed is inversely proportional to density of the medium. Galileo disposed of that easily, by noting that some bodies fall in air but rise in water. His whole argument throws light on the difference between the old natural philosophy and a new science devoid of reverence for casual reasoning of the form 'since weight is the cause of fall, the more weight, the more fall in the same time.'

It was also in the first 'day' that Galileo introduced the concept of one-to-one correspondence between members of infinite sets. Euclid excluded the infinite from mathematics by a simple axiom in Book I of his *Elements* – the whole is greater than a part. Greek philosophers had got into such troubles attempting to deal with the infinite that Greek mathematicians simply ruled it out. Galileo had to introduce it in order to derive the times-squared law of distance in fall, in the third 'day,' as will be indicated later. In the first 'day,' as preparation, he discussed paradoxes of the infinite and the cautions required in dealing with such problems mathematically. Among them he placed the 'soup-dish' paradox, which he had devised when advising his disciple Cavalieri about the treatment of limiting cases in the latter's 'geometry by indivisibles of continua' (published in 1635, but begun a dozen years earlier).

Near the end of the first 'day' Galileo stated the pendulum law, in connection with his own interesting theory of musical consonance and dissonance, and described experiments in which two or three pendulums returned periodically to synchronism. An arithmetical misstatement in that connection, and omission of the word 'inversely' in another statement, show that Galileo did not repeat these experiments late in life, if he had made them in his youth. But other statements of his in the first and third 'days' leave no doubt that Galileo had indeed observed long pendulums, probably mainly in 1602–4, and knew much more about them than has been supposed in the past. What has caused most of the trouble is that Galileo seems to have said several times in *Two New Sciences* that there is no difference in time for pendulums of the same length, whether the arc of swing is large or small, though in the earlier *Dialogue* he had granted insensible differences of time when two arcs of equal pendulums were greatly different. It has been natural to assume that in his old age, becoming more and more interested in mathematical conclusions, Galileo was less careful to mention small departures from actual physical observations. Attention to his exact words usually shows that not to have been the case.

First, it is important to remember that Galileo had no way to keep a pendulum swinging through the same arc repeatedly, and he made no attempt to do that. When he spoke of a great many oscillations, he referred to a pendulum started and then left to itself, damping down continually. Second, a phrase like 'equal among themselves,' rather than just the word 'equal,' was meant to say something specific, not easily said in a few words free from any ambiguity. And third, as was said in chapter 1, swing to the vertical only was of particular importance in Galileo's reflections about pendulum times. With those things in mind, a reading of the rest of the autobiographical passage from which a part was excerpted earlier will illustrate my point:

Ultimately I took two balls, one of lead and one of cork, ... attached to equal strings 8 or 10 feet long ... They sensibly showed that the heavy one kept time with the light one so well that not in a hundred oscillations, nor in a thousand, does it get ahead in time even by a moment ... The operation of the medium is also perceived; offering some impediment to motion, it diminishes the oscillations of the cork much more than those of the lead. But it does not make them more frequent, or less so; indeed, when the arcs passed by the cork were [i.e., had become] not more than five or six degrees, and those of the lead were [still] fifty or sixty, they were passed over in the same times.

Here the specification in degrees was for swing side to side, for even starting from the widest possible arc the lead bob will not still be swinging as much as 50° to the vertical by the time the cork bob has damped down to 5°, though by the time the lead is swinging 25° to the vertical, the cork will still swing about 3°. Now, when the arc to the vertical for the pendulum having the wider swing is no more than 25°, the difference of times for it and the other pendulum is not very great, and it keeps on diminishing. Sensible difference ceases well before a hundred swings – if swings still remain perceptible that long. From a hundred swings to a thousand swings, Galileo knew that no change could occur in either of the times. They would be truly equal for equal arcs, and sensibly equal whenever the wider of the two swings had damped down to 20° to the vertical.

This is also related to reconstruction of Galileo's thought about 'equivalent' pendulums in air and in a void, when he wrote the entries at the top of f. 189v1 early in 1604. Probably he already knew from experience that swings of less than 20° to the vertical had sensibly equal times, though of course he also knew that as long as swings are damping down there is necessarily some reduction in time of swing. When these things are kept in mind, the preconception that Galileo was straining to convince readers that perfect isochronism actually existed for pendulum swings can be set aside, and a clearer picture of his thought emerges:

If the lead pendulum [bob] is drawn, say, fifty degrees from the vertical and released, it passes beyond the vertical and runs almost another fifty, describing an arc of nearly one hundred degrees. Returning of itself, it describes another slightly smaller arc ... and after a great number of these it is finally reduced to rest. Each of those vibrations is made in equal times, as well that of ninety degrees as that of fifty, or ten, or four.

The last sentence seems to say absolutely that any size of arc does not affect time of swing, but that need not be meant. The plural 'times' is a bit unusual as applied to 'each'; possibly the thought concerned full swing from side to side and was that a downswing took the same time as the next upswing, or insensibly less, though the latter has a bit less amplitude than the former. Galileo's thinking on the matter becomes a bit more definite when Sagredo summarizes it on the next page:

... Nobody says that the speed employed in the arc of sixty [degrees] is equal to that consumed in the arc of fifty, nor this speed to that in the arc of thirty, and so on. The speeds are always less in the smaller arcs, which we deduce by seeing with our own eyes that the same body spends as much time in passing the large arc ... as in passing the smaller ...

Here the emphasis is on continual, nearly continuous change of speeds, arc by arc, in which the distance swung does differ from one arc to the next, though with damping down the difference becomes ever less. The conclusion, that the times are ever the same, followed only from the sensible evidence; hence Sagredo's final remark, 'and in sum, that all arcs are passed in equal times,' implied 'sensibly equal.' The discussion then passed on to the effect of air on motions, surface-volume ratios, and to speeds greater than can be acquired by fall through any medium, however long such fall may continue. To such speeds Galileo gave the name 'preternatural.'

Galileo knew that in continuous change, parts exist within which one cannot sensibly distinguish any difference between the values of the quantity at beginning and end, though a difference must certainly exist. The time of swing through 90° is sensibly different from the time through 50°, and that, again, from time through 20°, but the latter ceases to be sensibly different from time through 10°, or 5°. During damping down of swings, any differences of times after 20° of amplitude has been reached may be insensible. That is all Galileo probably knew, and beyond that he could not speak with assurance. How cautiously he dealt with such limitations will be seen later, when we consider the postulate he adopted for his new science of natural straight motions in the third 'day.'

Limitations of his previous sensible evidence probably led Galileo to one further pendulum experiment late in life, when a very interesting pendulum phenomenon was introduced in the fourth 'day.' It seems to many critics to prove that Galileo's claimed experiments were fictitious, a notion that arose through the unfortunate choice of one word on Galileo's part. Two equal pendulums about ten feet long were started simultaneously through very different arcs, and the numbers of swings were counted aloud by two persons. Galileo said that the numbers called out would not differ after tens and hundreds of swings; or rather, would differ by one. That is quite true; after about thirty swings, the counts would differ by one. Thereafter, no further change would occur because both swings will have arrived at very small arcs. But Galileo used the word *punto* for a count, in the sense of one point in a game – wrongly interpreted as meaning punctual agreement throughout.[1]

This suffices for statements about pendulums in *Two New Sciences*, except for some in the third 'day' and the postulate adopted there. In the second 'day,' as said before, we have the theorems on strength of materials, for which Galileo assumed the uniform distribution of the power of cohesion over the cross section, and considered the virtual lever formed by a beam that projected from a stone wall. He worked only in ratios, assuming the same material for beams differing only in dimensions. At the outset Galileo gave his own proof of the lever law, which is very interesting. It differs from the two-part Archimedean proof, in which the commensurable case was proved first, and then the case of incommensurable lengths. Galileo needed only the first of the Archimedean postulates, and the general theory of proportion from Euclid, Book v. Also, he cut a material prism instead of spacing equal weights equally along a mathematical line. He had already formulated this proof of the lever law in his 1601–2 *Mechanics*.

To the theorems on solid beams Galileo added his analysis of strengths of hollow tubes, and commented on maximizing ratios of strength to weight. Hollow bones of birds and hollow spears served as useful examples. Beams to support the deck of a ship, shaped to reduce weight without loss of strength, suggested his pioneering analysis of the beam that would break under a given loading at any point, first mentioned in the letter of February 1609 to Antonio de' Medici.

Tensile strength, as distinguished from resistance against transverse fracture, was touched on in the first 'day,' where Gali-

---

1   A Latin translation of the book, probably edited by Galileo himself, leaves no doubt of his meaning. Two early English translations were literal and not misleading, though a bit puzzling. Except for one Spanish version, all modern translations were simply wrong.

leo related it to the phenomenon of objects breaking under their own weight. In this connection he mentioned the failure of lift-pumps when the column of water was longer than about thirty feet. Galileo dealt with this phenomenon by analogy with a rope or rod, fixed at the upper end, which will part when it exceeds a certain length. Also in the first 'day' he dealt with the weight of air and described two ways in which he had measured that. Galileo never connected the weight of air with the phenomenon.

From the foregoing summary it can be seen that Galileo was a versatile and resourceful physicist, quite apart from the 'new science of motion,' for which his *Two New Sciences* is principally remembered today. Unlike the traditional treatises on motion, Galileo's was organized in the style of Euclid and Archimedes, with sequential theorems and problems preceded only by neces-sary definitions and postulates. The third 'day' is devoted mainly to natural motion, especially uniformly accelerated straight motion, and the fourth to projectile motions.

A single definition and a single postulate served for the thirty-eight principal propositions of the third 'day.'[2] Uniform accelera-tion was defined as that in which equal increments of speed accompany equal increments of time. Galileo postulated that the speeds acquired along planes of different slopes are equal when their vertical heights are the same. After the book was published, a number of friends told Galileo that his postulate was not easily assumed and needed better support. He then dictated to Viviani a dynamic lemma from which the proposition could be proved as an additional theorem, included in posthumous editions. This unconvincing attempt to invoke dynamic considerations without a mathematical definition of 'force' was carried out by reversion to an approach that Galileo had adopted and then discarded more than once during the years 1604–8.

Galileo's postulate, which his friends believed should be capa-ble of proof, was stated thus:

I assume that the degrees of speed acquired by the same movable [body]

---

2 Those were preceded by six theorems on uniform motion, with their own definition and axioms.

over different inclinations of planes are equal whenever the [vertical] heights of those planes are equal.

In support of the assumption Galileo first offered the case of a ball rolling down planes perfectly hard and smooth, in the absence of any material impediments. The fact that he included descent of a ball through the vertical as if it were the plane of maximum slope suggests that Galileo remained unaware that speed acquired by rolling differs from that acquired by fall. Yet that is not absolutely certain, since he did not speak of 'rolling' as such, but only of 'descending,' and in the absence of any friction whatever, a ball would not roll down a perfect plane, but would glide down. By excluding all material impediments to motion, Galileo's first example in support of his postulate was technically correct.[3]

To this support for his postulate, which Galileo had his spokesman call probable only, he added an experimental example that was not restricted to straight descent. A pendulum swinging an inch or two clear of a wall was affected at the vertical by a nail driven into the wall, its radius of swing thereafter being the string between the nail and the bob. The vertical height to which the bob swung upward remained the same for different places of the nail, until the radius became so short that the bob could not reach that height, and the string wound round the nail. The momentum acquired thus depended only on the vertical descent, and that alone determined the ensuing ascent. The following exchange ensued:

SAGREDO. The argument appears to me conclusive, and the experiment is so well adapted to verify the postulate that it may very well be worthy of being conceded as if it had been proved.

SALVIATI. I do not want any of us to assume more than need be, Sagredo; especially because we are going to make use of this assumption chiefly in motions made along straight surfaces, and not curved ones in which

---

3   The same reasoning had been put forth in the *Dialogue* and was accepted by the interlocutors as clear and evident.

acceleration proceeds by degrees [of speed] very different from those that we assume it to take when it proceeds in straight lines ... Hence let us take this for the present as a postulate, of which the absolute truth will be later established for us by our seeing that other conclusions, built on this hypothesis, do indeed correspond with and exactly conform to experience.

Galileo had been aware ever since 1602 that acceleration along circular arcs is different from acceleration along broken lines joining the same end-points. Within a year he had found the law of increase of speed from rest along a straight path, whether vertical or inclined, and by 1609 he understood its application to parabolic trajectories of projectiles by vector addition. In all his working papers there is only one conjecture about times along circular arcs. Their analysis had to await Newtonian dynamics.

Theorem I established, from the definition of uniformly accelerated motion, that speed acquired in fall from rest is double the uniform speed that would carry the body the same distance in the same time as that of the fall. In that form, the proof was probably the last proposition to have been written. It depended on Galileo's having put times and speeds during fall into one-to-one correspondence, a new mathematical tool of enormous power. Theorem II, the times-squared law of distances in fall from rest, followed immediately. In the discussion that followed it, Galileo described his apparatus and procedure for timing brief motions, as mentioned in chapter 1 above. Theorem III then established the proportionality of times to length and height of plane for inclined descent and straight fall. The attempted dynamic justification of Galileo's postulate was placed between these two theorems.

For Galileo's rigorous proofs of theorems and solutions of problems concerning motion in the third 'day' – of which some are quite complex – the inertial concept was not required. Its role for Newton was to establish a criterion for the presence of impressed force; for Galileo, all that was required was the concept of impressed motion. What he said about that is found in the scholium previously mentioned, in which he expanded on a note he had written in 1618:

Note that if motion along an incline is accelerated *in infinitum*, it seems one might demonstrate that along the horizontal, [motion] must extend equably [i.e., uniformly] *in infinitum*; now it is also clear that if equable, it will also be infinite.

In the scholium we find everything necessary for a complete understanding of Galileo's mature kinematics, and of the statements in his *Dialogue* that have induced historians to impute to him the meaningless and self-contradictory notion for which they have coined the phrase 'circular inertia,' as if he (or anyone else) had believed it possible for an unconstrained body to be so set in motion as to remain forever revolving uniformly around some center external to it. No one entertaining any such fantasy is likely to have written the following:

It may also be noted that whatever degree of speed is found in the movable [body], this is by its nature indelibly impressed on it when external causes of acceleration or retardation are removed, which occurs only on the horizontal plane ... From this it likewise follows that motion in the horizontal is also eternal, since if it is indeed uniform it is not weakened or remitted, much less removed.

Furthermore, one must consider the existing degree of speed acquired by the movable [body] in natural descent to be naturally indelible and eternal; but if after descent along a declining plane it is diverted through another plane upward, a cause of retardation presents itself there, for on such a plane the same body would naturally descend. Wherefore a certain mixture of contrary tendencies arises – that of the degree of speed acquired in the preceding descent, which by itself would carry the body away uniformly *in infinitum*, and [that of] a natural propensity to downward motion according to the same ratio of acceleration in which it is always moved. Whence it is seen to be quite reasonable if, in inquiring what events take place when a body is diverted through some rise after descent through some inclined plane, we assume that the maximum degree acquired in descent is in itself perpetually kept on the ascending plane; but in the ascent there supervenes the natural tendency downward – that is, to a motion from rest accelerated in the ratio always assumed.

The concept that a motion impressed in a heavy body is indelibly impressed had been applied in the *Dialogue* to explain why a weight dropped from the mast of a ship under sail falls to the foot of the mast, just as it falls when the ship is at anchor. Similarly, a bird resting on a rotating earth retains that impressed motion when it rises into the air, and beats its wings not to keep up with the moving earth, but to move with respect to that. To say that 'indelibly impressed *motion*' is evidence that Galileo believed in 'circular inertia' or any other kind of impressed *force* is not to try to understand his physics, but rather to show him incapable of thinking as a physicist at all, and only poorly as a metaphysician.

The proposition to which the foregoing scholium was attached stated a problem rather than a theorem:

Given the space run through in any time along the vertical from rest, to divert a plane from the lower end of that space upon which, after fall in the vertical and in equal time [to that of the fall], a given space is traversed that shall be more than double, but less than triple, the space run through in the vertical.

A preceding theorem asserted that after fall for a given time from rest, the distance traversed along any inclined plane during equal time will be more than double but less than triple the original distance during fall. Galileo had known since 1603 that along the horizontal, double the distance fallen to it would be passed in another equal time, and his law of fall in 1604 said that continuing in the vertical, triple the distance was passed in another equal time. This interesting appearance of the first two numbers after unity as the lower and upper limits of distance run along planes after an initial fall received no discussion or comment in Galileo's book. Had he thought like a metaphysician instead of a physicist, and especially if he had been a Platonist in any sense, he could hardly have resisted some numerological speculation of the kind enjoyed by Kepler. In a discussion at the end of Galileo's chord theorem, discussed previously, he placed such speculations in the mouths of the two other interlocutors and had his own spokesman say:

Such profound contemplations belong to doctrines much higher than ours, and we must be content to remain the less worthy artificers who discover and extract from quarries that marble in which industrious sculptors later cause marvelous figures to appear that were lying hidden beneath those rough and formless exteriors.

Such was Galileo's attitude toward science on the one hand, and philosophy and theology on the other hand. Near the opening of the third 'day' he had expressed it even more pointedly:

The present does not seem to me to be an opportune time to enter into the investigation of the cause of the acceleration of natural motion [of fall], about which various philosophers have produced various opinions, some of them reducing this to approach toward the center [of the universe]; others, to the presence of successively fewer parts of the medium to be divided, and others to a certain extrusion by the surrounding medium, which, in rejoining itself behind the [falling] body, goes pressing and continually pushing it out. Such fantasies, and others like them, would have to be examined and resolved, with little gain. It suffices ... to investigate and demonstrate some attributes of a motion so accelerated, whatever be the cause of its acceleration, that its speed goes increasing, after its departure from rest, [so] that in equal times, equal additions of speed are made.

The fourth 'day' presented theorems on projectile motion, beginning from the case of a body moving uniformly along a level plane and falling freely from its end. Galileo's mathematical derivation of the semi-parabolic shape of its trajectory brought forth numerous objections related to the practical applicability of his purely mathematical assumptions, raised by Sagredo as the spokesman for men of common sense, and also by Simplicio as the spokesman for Aristotelian philosophers. Speaking for

Galileo, Salviati replied by accepting all the difficulties they raised, adding others, and only then explaining how science proceeds in these matters:

No firm science can be given of such events of heaviness, speed, and shape, which are variable in infinitely many ways. Hence to deal with such matters scientifically, it is necessary to abstract from them. We must find and demonstrate conclusions abstracted from the impediments, in order to make use of them in practice under those limitations that experience will teach us.

At the time, this was truly a new view of science. That it is now taken for granted is a tribute to Galileo's pioneering work. His was a science of successive approximations, in which experience played an essential role at every stage of its progress, and no end to its progress was envisioned. Nothing of that kind had ever been proposed under the name of science as the domain of natural philosophy.

CHAPTER FIFTEEN

From Galileo
to Newton

In my reconstruction of Galileo's thought as a physicist, he abandoned the concept of an 'impressed force' as a cause of motion in 1598 and never returned to it. Except for his writings on motion before he moved to Padua, they are all kinematic, or at most kinetic, in conception. The concept of force, or of energy, remained undefined, as shown by Galileo's treatise on the force of impact (or of percussion as he called it). Written in his last years and intended for inclusion in *Two New Sciences*, that was withheld and never printed until 1718, long after his death.

So far as Galileo was concerned, force or energy was simply incommensurable with motion as such. Having no measure of force, he was content to discuss it only in general terms without trying to reduce it to a science. Force remained for him an indefinable term, as did energy, which Galileo frequently coupled with force, writing as if the two were synonymous.

The word *momento* was used interchangeably for static moment as well as for momentum in Galileo's writings. 'Moment of speed' was often identified with *impeto* by Galileo after he had freed the latter word from its medieval implication of an impressed force, and redefined it in the *Dialogue* to mean simply speed of a heavy body, however acquired. Galileo distinguished two components of *momento*, using the phrase 'moment of speed' for the component attributable solely to speed, and 'moment of heaviness' for the component due solely to the tendency to motion that could be measured by the weight of the

222 / Galileo: Pioneer Scientist

body.[1] We could represent Galileo's thought symbolically by writing M_v for 'moment of speed' and M_g for 'moment of heaviness,' his word for speed being *velocità* and for heaviness being *gravità*. Since he had measures for speed and for heaviness, the concept of 'moment' sufficed for his science of motion. Having no measure for force, he did not advance to dynamics from kinematics, although in some working papers he did attempt to ground his science of motion in *impeto* or in 'moment of heaviness' from time to time during the years 1604–8. Near the end of his life, at the request of friends, he dictated a passage for insertion into his last book that could replace his one postulate for those who preferred an argument based on a dynamic concept. But he did not remove the postulate asserting conservation of motion.

It was Newton who created the science of dynamics, though (unfortunately for him) he did not give it a distinctive name. Leibniz coined the word *dynamique*, and Newton was much annoyed when Leibniz treated this science as a creation of his own. In private papers Newton told its history as he saw it, and there he credited Galileo and Huygens alone as his predecessors in it.

Conspicuously absent from Newton's account of the history of dynamics was René Descartes, to whom it is not uncommon now to credit the inertial law. Newton's *wording* of his first law of motion is indeed traceable to wording that Descartes had used in his *Principia Philosophiae*, but the conceptual evolution of the inertial principle took more than words. For Newton, inertia was linked inseparably with force by his second law, which completed the statement begun in his first law, by defining the action of 'impressed force' in a manner that was very remote from anything that had been said by Descartes. In the very different Cartesian doctrine of conservation of motion, for example, what Descartes said flatly contradicted Galileo's law of fall, which Leibniz pointed out as 'the memorable error of Descartes.' Newton cannot have

1   It is essential to remember that the word 'gravity' meant heaviness, not weight, though heaviness was measured by weight and was accordingly proportional to weight. Newton pointed out that it was an abuse of language to identify things in nature with their measures.

failed to perceive that contradiction, whence his debt to Descartes was no more than literally verbal when he worded the inertial law as printed in his *Principia* of 1687.

Huygens was an acknowledged predecessor in dynamics for his having originated the concept of centrifugal force before Newton had arrived at it.[2] The related concept of centripetal force was Newton's own, and was essential to his origination of dynamics. Galileo was acknowledged for his law of fall, basic to Newtonian dynamics – not because Galileo brought into it the concept of force, for he did not, but because he introduced the analysis of uniformly accelerated motion, on which Newton's definition of force ultimately depended. In fact, Newton mistakenly credited Galileo with having discovered[3] the law of fall by anticipating the laws of inertia and of force, writing in his *Principia*:

By the first two Laws and the first two Corollaries, *Galileo* discovered that the descent of bodies varied in the squared ratio of the times and that the motion of projectiles was in the curve of a parabola; experience agreeing with both, unless so far as these motions are a little retarded by the resistance of the air.[4]

It was explained in chapter 1 how Galileo did discover the times-squared law, and in chapter 6 how he found the parabolic trajectory. In neither case did he apply either the inertial law or the force law. His two discoveries were based exclusively in careful

2   There is no trace of the notion of centrifugal force in what Galileo wrote in his *Dialogue* about the hurling of a body from a rapidly rotating wheel. His treatment of that phenomenon did include the word 'force' (*virtù*), but he applied it to an action centripetally, and his argument was entirely geometrical. It is quite possible that Newton, who read the *Dialogue* in English translation in 1666, was inspired by that to consider centripetal force, though that phrase was original with him.

3   The word Newton used was not the usual Latin for 'discovered,' but rather 'found out,' perhaps in the sense of his having made sure that the law was not merely empirical. Newton can hardly have imagined that Galileo had made no measurements; indeed, in 1666 he adopted figures from the *Dialogue* for his own first calculations of distances and times in fall. Galileo's figures were intended only as a round-number example and were not measured, but Newton believed they had been.

4   Scholium to the three laws of motion; Isaac Newton, *Mathematical Principles of Natural Philosophy* (Berkeley 1947), 21

measurements of motions – as I think it not unlikely for Newton to have surmised, despite what he said above. For he knew very well that both those dynamic laws were of his own devising, and not without many painstaking revisions from his original idea of presenting them as hypotheses, not laws.

It has long seemed curious that Newton, whose *hypotheses non fingo*[5] is a famous maxim in science, should ever have designated his three celebrated laws of motion 'hypotheses.' Why he did that will become clear from what follows.

In the third edition[6] of his *Principia* Newton added a long comment, after the passage crediting Galileo, which shows how Galileo's kinematics is related to Newton's dynamics. Below, I present this comment divided left and right by a space; ignoring the space and simply reading from left to right, the words are given exactly as translated from the Latin by Motte and Cajori:

When a body is falling,	the uniform force of
its gravity, acting equally,	impresses
in equal intervals of time	equal forces
upon that body,	and therefore
generates equal velocities;	
and in the whole time	impresses a whole force, and

generates a whole velocity proportional to the time.

Next, reading only the left-hand column, from top to bottom, what we have is a perfectly grammatical statement, devoid of any allusion to forces, of the kind that Galileo would have written on the basis of his definition of uniform acceleration in *Two New Sciences*. Newton's scholium (similarly divided) continued:

And the spaces described in proportional

5   What Newton said was that he 'feigned no hypotheses' concerning the cause of gravitation, having been unable to deduce one from the phenomena themselves. He intensely disliked the free use of arbitrary hypotheses by Descartes, but Newton by no means gave up the judicious use of reasoning ex hypothesi in science.
6   Meanwhile, the essay on dynamics by Leibniz had appeared, to the annoyance of Newton as previously mentioned.

times are as the products of the velocities
and the times; that is, as the squares of
the times. And when a body is thrown
upwards, its uniform gravity   impresses forces and
reduces velocities proportional[ly] to the times.

From the left-hand column it is evident that 'impression of a
force' was a superfluous concept so far as Galileo's kinematics
was concerned. It was *introduction* of that concept which immeasurably extended and advanced physics at Newton's hands. I see
no reason to doubt that Newton was perfectly aware of all this
when (for simplicity) he gave to Galileo more credit than was
literally due. A simpler alternative was to ignore Galileo's contribution entirely, but Sir Isaac would not do that. As he once said,
'If I have seen farther, it is because I stood on the shoulders of
giants.' Had he named them, he would very likely have included
Galileo. If Galileo had lived to read the *Principia*, he would certainly have regarded Newton as having fulfilled his prediction in
Two New Sciences that, by that book, 'there will be opened a
gateway and a road to a large and excellent science, of which
these labors of ours shall be the elements, [one] into which minds
more piercing than mine shall penetrate to recesses still deeper.'
A considerable amount of the debate over Galileo's ignorance
of the inertial concept, and Newton's supposedly profound debt
to Descartes for it, not only ignores the inseparable connection
between Newton's first and second laws of motion but also usually oversimplifies Newton's inertial law itself. It is very common,
but quite inaccurate, to say that Newton's inertial law asserted
that a body continues uniformly in a straight line when he wrote:

Every body continues in its state of rest, or of uniform motion in a right
line, unless it is compelled to change that state by forces impressed
upon it.

The qualification placed at the end, usually omitted for the
sake of brevity in the historical debates mentioned above, was
physically meaningless until Newton had defined the action of

the forces impressed, for whose presence he had thus far supplied only the criterion. The required definition was supplied next, in Newton's second law of motion:

The change of motion is proportional to the motive force impressed; and is made in the direction of the right line in which that force is impressed.

Uniform motion in a straight line was accordingly Newton's criterion for the total absence of any impressed forces whatever, if any such motion – which we call 'inertial motion' – ever happened to exist and to be observed, something *not* asserted. It was this definition of their magnitude and direction that created the science of forces; that is, dynamics. For Descartes, uniform straight motion was a criterion not of the absence of anything, but of the presence and immutability of God, known intuitively, and of the inability of anything to change itself, by metaphysical principles. Newton doubtless endorsed those concepts, but they were of no help to him in creating the science of dynamics.

As for Galileo's ignorance of the inertial law, that is no more than a pejorative way of saying that Galileo did not create the science of dynamics. Neither did anyone else before Newton. To credit Descartes with anything useful to dynamics is mistaken. It is indeed surprising to see how seldom Descartes even employed the word 'force.' The same is true of Galileo, once he had got over his youthful acceptance of 'impetus.' He did not propose a force to explain conservation of motion, or impression of a force to explain change of speed. That fact astonishes Galileo's critics. No one is astonished that Huygens did not state the inertial principle, even though he had read Descartes with care and admiration. How Newton is supposed to have perceived something there that Huygens missed has not been explained by those who believe Descartes to have anticipated Newton in conceiving the principle of inertia. What Newton perceived was conceptual, not merely verbal, and the words Descartes wrote fell short of it.

Consider matters the other way round, asking how, of all people, Newton could ever have possibly embraced the idea that a uniform straight motion could exist at all. By Newton's law of universal gravitation, no body in the universe could possibly move forever uniformly and in a straight line. Gravitational force would inevitably be impressed upon it at some place in its infinite progress, by some other body, and then its motion could no longer remain uniform and rectilinear. Newton's qualification in wording his inertial law, far from asserting uniform straight motion, allowed for the action of universal gravitation.

It should now be easy to see why Newton had at one time called his laws of motion not 'laws,' but 'hypotheses.' It was a hypothesis, not verifiable by observation, that a body in uniform straight motion would continue forever in that motion. It would remain a hypothesis after adding 'unless acted on by an impressed force.' But when the effect of such a force was specified both as to direction and amount of change in the hypothetical motion, the statement became directly verifiable by observations and measurements. Thus verified, the statement became an assertion about natural phenomena; that is, a law of nature.

Because Galileo and Huygens did not conceive of gravitation as universal, their physics required no inertial principle, but only conversation of received motion, which might perhaps be called a 'local inertia.' Descartes, having declared that a body would naturally move forever in a straight line, provided against its ever doing so by filling space with 'subtle matter,' whirling in vortices to divert it from such a path, destroying rather than conserving its natural motion. Newton's physics was the first to require the *vis inertiae*, or 'inertial force.' By now, even matter has been reduced in principle to the actions of forces, a physics inconceivable before Newtonian dynamics, and long delayed after his original formulation of the force concept.

What took the place of the inertial concept in Galileo's kinematics, as previously remarked, was conservation of impressed motion; that is, of speed. Behind that lay his understanding of relativity of motion, physical as well as optical; that is, his understanding that motion is measured with respect to something assumed to be at rest for the purpose in hand. Huygens took up the Galilean concept of relativity of motion and developed it further, geometrically, with great ingenuity. Newton discussed relativity not in terms of motion, or of speed, but of space and of time, emphasizing that all measurements are necessarily not absolutes, but comparisons, so that measures of distance and of time inform us only of relations and not of the things measured.[7]

Descartes also insisted on the relativity of motion, but more to avoid conflict with theologians than as something to be applied in all physical investigations. Motion or rest of the earth became alternative ways of describing a physical situation, a merely verbal and arbitrary exercise, so that no conclusion about motion or rest of the earth need be asserted by Cartesians. What Descartes did assert was that the quantity of motion in the universe remains forever unchanged, a conservation principle of no concern to theologians. The manner in which Descartes treated 'quantity of motion,' however, entailed his 'memorable error' of contradicting Galileo's law of falling bodies, as pointed out by Leibniz.

That contradiction cannot have been overlooked by Newton. It alone would have prevented him from listing Descartes among the giants on whose shoulders he had stood, at any rate so far as the physics of motion was concerned. Even in the physics of light, to which Descartes contributed scientific analysis of the rainbow, he may not have seemed a giant to Newton. But in mathematics there is no question that Descartes was a giant on whose shoulders Newton stood. Systematic application of algebra to geometry by Descartes was a monumental contribution to the swift growth of mathematical physics. Newton adopted Cartesian analytic geometry from the first, but in later years he expressed regret that he had not paid closer attention to Euclid as a young student. Because he did not further explain that remark, and it may seem odd to scientists, it deserves brief comment here.

Newton cannot have been alluding to the strictly geometrical books of Euclid's *Elements*, for those included no theorems that could not be dealt with more efficiently by applying Cartesian analytic geometry. Still less can Newton have meant the so-called

---

7 'Those violate the accuracy of language, which ought to be kept precise, who interpret these words [time, space, place, and motion] for the measured quantities. Nor do those less defile the purity of mathematical and philosophical [i.e., physical] truths who confound real quantities with their relations and sensible measures,' (*Principia*, scholium to the definitions).

arithmetical books of Euclid, which might better be called the 'number-theoretic' books, mainly Books VII – IX, and in a sense also Book X. Newton must have had in mind especially Euclid's general theory of ratios and proportionality in Book V. For the essential difference between his 'method of fluxions' and the infinitesimal calculus of Leibniz was Newton's emphasis on the limit approached by a ratio, as against the indefinitely small terms of a ratio that are called 'actual infinitesimals.'

Here, parenthetically, I should like to comment on the unseemly dispute that arose between followers of Newton and the adherents of Leibniz over priority of invention of the calculus, into which Newton himself was drawn, behaving no more fairly than his adversaries. In view of the above conceptual gap, it is impossible that either Leibniz or Newton derived the slightest benefit from learning, openly or surreptitiously, about the work of the other in this invaluable field of mathematics. The ratio-approach of Newton, essentially geometrical in spirit, differed entirely from the approach through vanishingly small quantities, essentially algebraic and numerical, of Leibniz. The difference is analogous to the unbridgeable gap between the continuous and the discrete recognized – or at any rate assumed – by ancient Greek mathematicians, inducing Euclid to deal with the theory of proportion separately for numbers (in Book VII) and for continuous magnitudes (in Book V).

The mistaken notion that Galileo's mature physics in his last two books somehow called for an inertial concept that he never stated is responsible for the modern imputation to him of belief in 'circular inertia.' Feeling that something should be present that is not to be found in Galileo's books, historians invented a notion discreditable to him as a physicist. Nothing useful could be gained in that way; no light could be thrown on Galileo's creditable physics by supposing him deficient in good sense. On the contrary, a good deal was lost by this historical gambit, which made it appear that Newton could have gained little or nothing of use to him by reading Galileo's books. That conviction, in turn, compelled historians to look elsewhere for a source of Newton's physics, which they assigned to Descartes. In that way they may have missed something important that Newton got from reading

Galileo, just as they wrongly credited Descartes with a Newtonian concept very different from his.

Newton read the *Dialogue* during his student days, when the university was closed in 1665–6 because of the Great Plague. A page of his working papers bears entries which show that he used the 1661 English translation by Thomas Salusbury. Salusbury's translation of *Two New Sciences* was printed in 1665, but very few copies survived the Great Fire of London. Specialists doubt that Newton read *Two New Sciences*, but agree that he read the *Dialogue* at a time when there were few books containing information about physics other than the standard Aristotelian variety.[8] For that reason, and because Newton was then a student to whom Galileo was a scientific author of high reputation, it is safe to assume that he read the *Dialogue* not to detect and correct or expose errors in it, as many historians do now, but to gain as much information from it as he could by thoughtful reading.

The following interchange between Galileo's own spokesman and the representative of philosophers could hardly have escaped notice by Newton, and would certainly have come to him as new:

SALVIATI. I did not say that the earth has neither an external nor an internal principle of moving circularly; I say that I do not know which of the two it has. My not knowing that does not have the power to remove it. But if this author [an anti-Copernican] knows by which [kind of] principle other world bodies are moved around, as they certainly are moved, then I say that what makes the earth move is a thing similar to whatever moves Mars and Jupiter, and which he believes also moves the stellar sphere. If he will advise me as to the motive power of one of those movable bodies, I promise I can tell him what makes the earth move [around the sun]. Moreover, I shall do the same if he can teach me what it is that moves earthly things downward.

SIMPLICIO. The cause of this effect is well known; everyone is aware that it is gravity.

---

8 Newton's remark about Galileo's discovery of the parabolic trajectory, and another passage commenting on this as 'Galileo's theorem,' incline me to believe that Newton had read *Two New Sciences* also, but here I will speak only of the *Dialogue*.

SALVIATI. You are mistaken, Simplicio; what you ought to say is that everyone knows it is called 'gravity.' What I am asking for is not the name of the thing, but its essence, of which essence you know not a bit more than you know about the essence of whatever moves stars around. I except the name that has been attached to it and has become a familiar household word through the daily experience we have of it. But we do not really understand what principle or what force it is that moves stones downward, any more than we understand what moves them upward after they leave the thrower's hand, or what moves the moon around.

Galileo's remark that full understanding of gravity would enable him to say what makes the earth circle the sun, and his linking the fall of stones with the moon's circling the earth, may have started Newton on the train of thought that culminated in his greatest discovery, universal gravitation. I do not say that it did, but the possibility is heightened by Newton's telling a friend late in life that the fall of an apple started him thinking of the moon's orbiting around the earth. Such an association would be unlikely unless something else had come to Newton's attention recently that linked the two phenomena. It is hard to think what that might have been, other than Galileo's unusual promise in the *Dialogue*.

Two pages later came this interchange:

SALVIATI. ... So that if the terrestrial globe were pierced by a hole passing through its center, a cannon ball dropped through this and moved by its natural and intrinsic principle would be taken to the center, and all this motion would be spontaneously made and by an intrinsic principle [of motion]. Is that right?

SIMPLICIO. I take that to be certain.

SALVIATI. But having arrived at the center, do you believe that it would pass on beyond, or that it would immediately stop its motion there?

SIMPLICIO. I think it would keep on going a long way.

SALVIATI. Now, wouldn't this motion beyond the center be upward, and according to what you have said, be preternatural and constrained? But upon what other principle will you make it depend, apart from the very one which brought the ball to the center and which you have just called intrinsic and natural? Let me see you find an external mover who shall overtake it again to throw it upward.

About forty pages earlier, Galileo had discussed the notion that swift rotation of the earth would fling off heavy bodies resting on it, a misapprehension introduced and inadequately refuted by Copernicus himself.[9] Assuming that if a body were flung off, it would move uniformly along a line tangent to the earth, and using Euclid's theorem that the 'mixed angle' between tangent and surface is less than any rectilineal angle, Galileo proved that no body can leave the earth by reason of any speed of its rotation, and that a body supposed to have left it in that way would return to the surface, not escape from it. The assumption was equivalent to Newtonian inertial motion. One cannot say that Newton got his inertial idea from Galileo's assumption, but it is possible. It is hardly possible that Newton obtained it from the unqualified assertion made by Descartes. Galileo's argument was mathematical only, from a physical assumption. No concept of force was used; it is clear from Galileo's diagram that his treatment of speeds of return was rooted in his law of uniformly accelerated fall.[10]

How much Newton got out of all this is debatable, but if we assume either that he read the entire second 'day,' or that he already knew the times-squared law of fall before reading the *Dialogue*, he could have got quite a lot. Ten pages after the discussion of projection by whirling came Galileo's discussion of objects attached to a wheel that are thrown off, having no tendency to approach its center. Newton's first working papers on

9 Copernicus ascribed this argument against daily rotation of the earth to Ptolemy, mistaking his meaning in the *Almagest*.
10 The 'triangle of speeds' in uniform acceleration was simply laid horizontally, instead of placed vertically in the usual way. At this point in the *Dialogue* Galileo had not yet mentioned the law of fall, so he did not invoke it, though it had probably led him to his no-projection proof in the first place.

centrifugal force were written during the same period of closing of Cambridge University when he was reading Galileo's *Dialogue*.

All these events may have been merely coincidental, but they suggest that Newton in fact obtained a great deal more from his reading of Galileo's work than is currently supposed. As that work becomes better understood in its own right, and not just as a defective attempt at the dynamics that Newton finally wrote, its relation to that later physics will likewise become clearer than it has been up to the present. One thing is already virtually certain – that Galileo's kinematics was almost the same as Newtonian kinematics. The passage cited earlier from the third edition of the *Principia*, crediting Galileo, shows that no conflict existed and suggests that Newton himself was well aware of that fact. That does not mean that Newton took anything from Galileo, or from anyone else. It means that a viable kinematics existed and awaited only someone to move on from it to a dynamics consistent with the law of falling bodies and with conserved motion, fruits of Galileo's patient and ingenious experimental investigations.

James Clerk-Maxwell opened a famous address a century ago with the remark that experiments in science had become almost exclusively measurements. In his specialty, electrodynamics, measurements had hardly been possible when he was born, but by the time he died they had become the most precise measurements of all. When Lavoisier was born, chemistry was hardly yet a science, for lack of precision in measurements; Lavoisier transformed it by his uses of the balance and pneumatic trough. When Galileo was born, two thousand years of physics had not resulted in even rough measurements of actual motions. It is a striking fact that the history of each science shows continuity back to its first use of measurement, before which it exhibits no ancestry but metaphysics. That explains why Galileo's science was stoutly opposed by nearly every philosopher of his time, he having made it as nearly free from metaphysics as he could. That was achieved by measurements, made as precisely as possible with the means available to Galileo or that he managed to devise.

Galileo used the word 'science' and distinguished that from philosophy in many places in his books. Newton, I believe, did not use the word 'science'; he used the phrase 'experimental philosophy' to designate the kind of physics that began with Galileo. It was a very apt phrase, much more descriptive than the now frequent 'experimental science,' already obsolescent when Clerk-Maxwell delivered his address identifying experiment with measurement in science.

It was Heinrich Hertz who best explained the role played by experience in physical science:

We form for ourselves images or symbols of external objects, and the form which we give them is such that the necessary consequents of the images in thought are always the images of the necessary consequents in nature of the things pictured. In order that the requirement may be satisfied, there must be a certain conformity between nature and thought. Experience teaches us that the requirement can be satisfied, and hence that such a conformity does exist.[11]

Galileo's philosophy of science is summed up in modern language by Hertz's statement. When the thought is mathematical, the images formed are of measurements in nature and their necessary consequents are other measurements. Hertz's word 'external' shows that to his epistemological statement we may safely add the remark with which P.A.M. Dirac is said to have usually opened his course in quantum mechanics:

The existence of an external world is assumed. That is all the metaphysics you will need for this course.

Such was Galileo's metaphysics and his epistemology. If more is required for a philosophy of science (as, for example, an ethics and an aesthetics), I leave it to philosophers to supply those for Galileo.

---

11 Heinrich Hertz, *Principles of Mechanics* (trans. Jones and Walley, London 1899), preface

CHAPTER SIXTEEN

Galilean Units Today

From the time of Newton to the present century, advancement of physics depended mainly on dynamic investigations. Study of motion for its own sake, pioneered by Galileo, gave way to study of forces when dynamics was created. Until 1905, when Einstein proposed his special theory of relativity, kinematics had fallen into relative neglect. The title of Einstein's paper hardly even hinted that both the concept of simultaneity and the measurement of time required fundamental revision, for that title was simply 'Electrodynamics of moved bodies.' Yet that paper inaugurated a renaissance of kinematics because Einstein re-examined elementary concepts at a highly advanced stage of physical knowledge.

Sir Arthur Eddington, one of the first to support Einstein's theory of relativity, perceiving the dependence of our 'physical constants' on units of measurement, startled most physicists only a quarter-century ago by deriving the number of electrons in the universe from the numerical values of accepted physical constants.

The full implications of units that are adopted for physical measurements may go unnoticed by those who introduce, define, and use them for special purposes. Once introduced, scientific units soon become standardized, after which it would be very bothersome to substitute others. Indeed, that would in most cases be pointless, because units of measure ultimately depend on conventions, making them in the last analysis arbitrary choices. Nevertheless, it is sometimes possible to derive non-arbitrary units from phenomena of the physical universe. That is what Galileo did in 1604, as you may see by reviewing chapter 1. As a result, Galileo pioneered one pair of units that is of astronomical interest and potential utility today, though he could apply it only to terrestrial uses.

The units most commonly used are those of length, time, and mass – now standardized in the metric system as the meter, the second, and the gram. The unit of mass was related to the unit of length by establishing the gram as the mass of 1 cc of water (at 4° centigrade). The units of time and length were left still unrelated, though the commission of scientists empowered to create the metric system did consider relating them by fixing the meter as the length of a seconds-pendulum. It does not matter that they rejected the proposal, since there is a better way to go about the linking of length with time by use of the pendulum.

The second of time is 1/24 of 1/3600 of the duration of one rotation of the earth on its axis, a time-interval very accurately known. But even though based in celestial phenomena, the second is an arbitrary unit of time. Its division by 24 survives from the 12 zodiacal signs of Babylonian astrology, while $3600 (= 60^2)$ uses a number-base abandoned long ago in favor of 10. To avoid any arbitrary division of time whatever when linking it with the division of length, the motion of a pendulum does offer a means. But the pendulum of real value is not the seconds-pendulum, or any 'time-unit-pendulum,' so to speak. The 'reciprocal-pendulum' is best, although it has never received the attention it deserves.

The standard seconds-pendulum is $g/\pi^2 = 0.99362138$ meter in length; the reciprocal length is 1.006419562 m. This reciprocal-pendulum, of length 1.006419562 m, has the interesting property of swinging from side to side in 1.006419562 seconds of time. A one-'meter' pendulum beating one second, which is what the metric commission considered and rejected, would of course equal its own reciprocal-pendulum, and nothing could be learned from it that was not already known in order to establish it. But any pendulum of length x which beats x time-units offers a length-unit not only commensurable, but numerically identical, with a certain measure of time, in units that are not arbitrary but fully determined by gravitational phenomena. Potentially it could be useful to make our measures of length and time 'gravitationally commensurable,' whether or not space and time themselves are commensurable. That can be done simply by a suitable choice of units of measurement.

The manner in which Galileo created his own time unit, the *tempo*, to fit gravitationally with his arbitrary length unit, resulted in his pioneering not only a new pair of units, but also (unknowingly) a gravitational constant of acceleration that remains valid anywhere that bodies fall and pendulums oscillate.

We know that bodies fall much less rapidly on the Moon than on Earth, so what was just said may seem impossible. Nevertheless there is a pair of units in which acceleration during fall is the same regardless of the strength of any 'gravitational field' as that strength is measured dynamically.

From the equations for distance acquired in fall from rest, $S = (g/2)t^2$, and for time of a pendulum swinging through a small arc to the vertical, $t = (\pi/2)\sqrt{(l/g)}$, it is easily seen that the ratio of *times* when $s = l$ is simply $\pi/2\sqrt{2}$, a purely mathematical ratio independent of units of time or of length. Likewise when t is the same for a fall and a pendulum, their ratio of *lengths* is $\pi^2/2^3$, no matter what units of measure are adopted.

The above ratio of times may be denoted $\sqrt{g} = 1.110720738$. I call this 'Galileo's constant,' because his $942/850$ ($= 1.108$) was the ratio he measured for the time of swing of a pendulum to the vertical over the time of fall from rest through its length. The physics that Galileo pioneered was a physics of relations that he was able to measure, objectively and as exactly as his apparatus permitted. In such a physics there is no hope of any dynamics, or of *force* in general, because Galileo could not measure it.

Our metric gravitational constant of acceleration near the earth's surface, $g = 9.80665$ meters per second-squared, is exact only at sea-level at latitude 45° (or gravitationally equivalent places). Strictly speaking it is not a constant, but varies with the distance of a falling body (or of a pendulum bob) from the center of mass of the earth. It is clear from the definition of 'fall,' which implies changing distance between the falling body and the center of the earth, that a *constant* of gravitational acceleration could literally exist only if it truly remained the same anywhere that bodies will fall and pendulums swing. Metric standard g cannot do that, it being a dynamic and local relation with respect to terrestrial gravitation. In contrast, $g = \pi^2/2^3$ is unaffected by local conditions and it may be applicable to the planets, taken one by one in relation to the Sun.

To see this possibility, let us consider γ as representing rates of gravitational acceleration in general, whether dynamic and local like g or kinematic and universal like g. Numerically, γ can have any value we wish to assign, except zero, for γ is completely determined by the units that are used for measuring lengths and times. When γ is put equal to $\pi^2/2^3 = g$, fall from rest timed by vertical swing of a pendulum of length l is gl, and the time of such a swing exceeds the time of fall through length l by the factor $\sqrt{g}$, in any units of time and length.[1] When t is the same for a given distance of fall from rest and the swing of a pendulum to the vertical through a small arc, length of that pendulum exceeds the given distance of fall by the factor g. The units of length and time that make $\gamma = g$ will be designated by λ for length and τ for time; that is, acceleration will be $g \; \lambda/\tau^2$. Those units will be called GU, for Galilean units, and they make lengths and times gravitationally commensurable entities[2] so far as the rate of acceleration is concerned.

There is another gravitational constant in metric units, now usually denoted by G and called 'big G,' that is universal and not local, but that is not the measure of a rate of gravitational acceleration. The mass-distance relations established by the law of universal gravitation discovered by Newton, holding true throughout the entire universe, determine G; and of course the rates of acceleration associated with gravitation in general are as many and varied as are masses and distances in the universe.

1 $\sqrt{g} = \pi/2\sqrt{2}$ should be called 'Galileo's constant,' because he had measured it very nearly before he recognized the pendulum law (though he never realized that π was involved at all). See S. Drake, 'Galileo's constant,' *Nuncius* 2:2 (1987), 41–52.

2 To say that measures of length and measures of time can be made commensurable carries no implication that space and time are also commensurable entities; as Newton remarked, it is an abuse of language to confuse things with their measures.

In chapter 1 it was seen how Galileo's units of length and time were linked in the course of finding the pendulum law. His *punto* was 0.94 mm and his *tempo* was 1/92 second. In the units λ and τ, which make gravitational acceleration g λ/τ^2, I calculate λ to be 0.942212 mm, and τ to be 1/91.88025 second, agreeing with Galileo's pair of units as nearly as those can be determined from his working papers (which is only to 2 significant figures). My calculations of the above values for λ and τ in metric units are not theoretical, but are based simply in the ratios of planetary data reported in readily accessible astronomical tables for the axes of orbits and the periods of revolution about the Sun.

In GU, the semimajor axis of Mercury's orbit is $61.608877 \times 10^{12}\lambda$. The mean distance of Mercury from the Sun is reported, to 3 significant figures, as being 57.9 million km. In GU that amounts to $61.45 \times 10^{12}\lambda$. Mercury's mean distance must be less than its semimajor axis, which by my calculation is 58.0488 million km. Of possible interest to astronomers would be knowing data in metric units more exactly than at present. Certainly it would be useful, in the precise determination of the masses of Sun and planets, to know the value of metric G to 8 places, now given only to 4 or 5. Evidence that GU can provide such information will be indicated below, with only such explanatory background as is appropriate to the theme of Galileo as a pioneer of modern physical science.

In this chapter, basic planetary data will be designated by initials of the names of planets, with m for Mercury and M for Mars. The initial alone denotes the semimajor axis of a planet. For my present purposes I shall use not AU, but Mercury units with m = 1; m − denotes the semiminor axis of Mercury, m′ its sidereal period (in the customary Earth days), and m″ represents the orbital eccentricity of Mercury. When any measures are given in GU, the initial letter is italicized; in GU, no planet is assigned unit value. In order that anyone may verify statements made below, basic data taken from standard tables in AU (E = 1) are converted in table 1 into mU, with m = 1. Planets beyond Jupiter are omitted, as discussion here will be chiefly devoted to Mercury, Venus, and Earth alone.

In mU, the semiminor axis m − is a logarithmic function of g

Table 1 (in mU)

Semimajor axis		Period in Earth days		Eccentricity of orbit		Semiminor axis	
m	1.000000000	m′	87.969256	m″	0.205614	m−	0.978633152
V	1.868595722	V′	224.700789	V″	0.006821	V	1.868552248
E	2.583320481	E′	365.25636	E″	0.016751	E	2.582958037
M	3.936183388	M′	686.979702	M″	0.093309	M	3.919001669
J	13.44051012	J′	4332.5879	J″	0.048254	J	13.424853223

(and of 10^7, the role of which factor in GU will be seen when those units are later used). In mU, the ensuing expression for m− is in exact agreement with the currently measured data, those being reported in ordinary tables to 10 significant figures:

$$m- = 1/\ln \ln \ln [10^7/\ln \ln \ln(10^7 g)]$$

It could, of course, be mere coincidence that exact agreement to 10 significant figures exists between the measured semiminor axis of Mercury and a function of $\pi^2/2^3$ and 10^7. But inasmuch as a gravitational meaning for g has first been assigned, and the above expression is not without symmetry, other non-Keplerian relations among measured planetary data seem worth looking for. From this one, the orbital eccentricity of Mercury can be exactly calculated to any desired number of places by textbook methods.

Here an objection arises, for it is known that the axes and eccentricities of planetary orbits vary secularly, and there are tables for those adjustments. That might not necessarily be the case if the semimajor axis of Mercury were always taken to be our unit of distance (whatever it might happen to be in millions of kilometers at any particular calendar time). As a start, we may regard the above formulation as supplying a 'focal' value, so to speak, from which Mercury's semiminor axis varies secularly but only within narrow limits, while its measured orbital eccentricity automatically follows suit.

There are two reasons for starting thus. One is that focal values can also be found for the ratios E/m and E′/m′; for V′,[3] the sidereal

3   The case of V′ will be deferred until measures in GU are discussed, below.

period of Venus, and for the orbital eccentricity of Venus, V''. The other reason is that V'' can be found in two ways, one having the form $V'' = f(g, 2, g')$. Calculation of the focal value of V'' as such a function requires no mathematical operations beyond those of squaring, doubling, and point-shifting:

$$V'' = 2/10(1 + g'/2\sqrt{g})^2 = 0.0068217807$$

where $g' = 9.806768368$, the calculated kinematic g in GU. Now, by the use of logarithms, nearly the same result is obtained from the sidereal period of Venus in Earth days; that is, $V'' = f(V')$:

$$V'' = 2/10(\ln V')^2 = 0.006821345$$

It might appear that $\ln V' = 1 + g'/2\sqrt{g}$ (or $\ln[V'/e] = g'/2\sqrt{g}$), and that accordingly the value of g' above may require some adjustment, because the astronomical measurement is certainly trustworthy to 8 significant figures. But every measurement has a particular date, whereas the focal value is by definition independent of time.

Other focal values from which certain planetary relations do not widely depart include the ratio E/m of the semimajor axes for Earth and Mercury. E/m is now measured as being 2.583320481. At first glance, this seems to be just a random collection of digits, but that is not the case. Any number >1, subjected repeatedly to the operations $\ln$, $+1$, $\ln$, $\times 10$, $\sqrt{}$, will converge to 2.583322768, which may be the focal value from which the ratio E/m varies but little. Possibly the ratio E/m reflects kinematic gravitational laws acting to focus its value at a/b, such that $10 \ln \ln e(a/b)$ remains very nearly $\sqrt{(a/b)}$ by continuously approaching the above value as a limit.

Now, by Kepler's third law, $(E/m)^3 = (E'/m')^2$, which, at the above convergency-value, would put E'/m' at 4.152101938 (currently it is 4.152091044). The convergent operations for E'/m' are $\log_{10}$, $1/x$, -1, $1/x$, $\sqrt{}$, $\ln$, $1/x$, with the convergency at 4.151172975. If either ratio reached its convergency-value, the other could not be simultaneously at its convergency-value – and an absolutely steady state of axes and periods for Mercury and Earth may be impossible, even without our taking dynamic interplanetary perturbations of motions into consideration.

These purely kinematic considerations call into question the cosmological assumption that *positions* of planets became fixed solely as the result of *dynamic* laws of gravitation. There can be no doubt that those laws are correct and exact, within the limits of accuracy of astronomical measurements. That, however, does not exclude the possibility that planets may occupy positions that are determined by relations ultimately expressible as $f(\pi, 2)$.[4]

In what has been said here to illustrate the appropriateness of GU for kinematic planetary investigations, the operations of doubling, squaring, and reciprocal create no puzzles concerning gravitational relations; but multiplication and division by 10, or by powers of 10, do appear out of place. How powers of 10 enter into the matter will not be discussed here beyond the remark that the symbol 10 in decimal notation stands for the number ten, but in any place-notation it is the point-shift symbol, and in the binary notation based on 2 it stands for two, so that shift of the point to right or left doubles or halves the magnitude that is represented.

It is a uniquely valuable property of GU that, when we put acceleration due to gravitation at g λ/τ^2, the square of the *time* in τ of any fall from rest, numerically, is the *length* of pendulum in λ which, by its swing through a small arc to the vertical, will exactly time fall from rest through *double* the original distance. That provides a critical test for verifying that astronomical or other physical measurements have been exactly represented in GU, for no such relationship exists except in GU-related units.

Recall now the seconds-pendulum. The τ-pendulum, having length g/π^2 λ $(= \lambda/8)$ would be universal – if there were such a thing as a *pendulum* about 0.12 mm long. Anywhere in the solar system, such a 'pendulum' would swing side-to-side in time τ – in principle. In fact, it could neither be anchored to a fixed support, nor swing, nor be directly timed experimentally.

Now, any pendulum whose swing can be timed is the unit-time pendulum at some place, in a designated set of units of our

---

4 It may seem that 10 should be included, but this point-shift symbol in any place-system of numeration stands for 2 in binary arithmetic; doubling and halving abound in natural laws, from which 10 is absent.

own selection, for the same reason that the seconds-pendulum will beat seconds only at latitude 45° and sea-level (or at gravitationally similar places elsewhere). Likewise, any pendulum situated where it happens to have the length that is the reciprocal-pendulum to the unit-time-pendulum for the length and time units chosen has the property of possessing as many length-units as its swing side-to-side has time-units, from what was said earlier. The relation 'as many x as y' is among the most valuable concepts that exist.[5] Gravitationally, it entails the relation of a square to a square root – operations which, along with π, suffice for measurement of time in GU by means of gravitational phenomena.[6]

Galileo's constant, $\sqrt{g} = \pi/2\sqrt{2}$, well illustrates one general rule holding for different constants of gravitational acceleration. In $1/\sqrt{g}$ seconds (= 0.9003163162 s), fall at g (= 9.80665 m/s^2) is 3.974485542 m. The ratio of g to $\mathbf{g} = 7.948971085$;[7] that is, $g/\mathbf{g} = 2 \times 3.974485542$ – and for any rate γ of acceleration, $\gamma/\mathbf{g}$ is the double of the distance fallen, at rate γ, in $1/\sqrt{\mathbf{g}}$ time-units of γ.

The distance fallen at acceleration $\mathbf{g}$ in the GU time of 1 second (= 91.88024932 τ) is 5,207.437917 λ ($\cong$ 4.9065 m). Twice the distance fallen in 1 second of time is 10,414.87583 λ. Letting $10^2\sqrt{(2G/\mathbf{g})}$ = double the fall in 1 second at rate $\mathbf{g}$ (which fall is timed by the pendulum of length $(s/\tau)^2\lambda) = 8,441.980215$ λ. Now, at rate $\mathbf{g}$, this pendulum swings to the vertical in 129.938295 $\tau = \sqrt{2} \times 91.88024932 = \sqrt{2}$ s. Fall at $\mathbf{g}$ in that time is 10,414.87563 λ, double the original fall; and converting the double-fall to meters yields the rate $g' = 9.806768368$ m/s^2, slightly above the standard metric rate of 9.80665 m/s^2. Next, putting

---

5 Galileo was the first to point this out when, in *Two New Sciences*, he introduced the use of one-to-one correspondence between members of infinite sets to derive the law of fall rigorously.

6 Historically, time *was* measured by gravitational phenomena – the motions of heavenly bodies, descent of water or of sand, and, from the time of Huygens, the swings of pendulums. Only in the present century have phenomena of light and of atoms been used.

7 The ratio $g'/\mathbf{g} = 7.94906703$, g' being the kinematic g. $(g' - 7\mathbf{g})/\mathbf{g} = 0.94906703 \cong \ln(E-/m)$ at $E-/m$ 2.583298396.

$\sqrt{(2G/g)} = 104.1487583$, $G = 6.678109968 \times 10^3$. In order of magnitude, G is roughly $10^{14}G$.

A second of time can be nearly expressed in GU by the sole operations of square, square root, and reciprocal,[8] with attention to point-shifts in our decimal notation. In GU, very nearly,

$$1 \text{ second} \cong 10^2(10/\pi^2)^{1/8} = 91.85097747 \ \tau.$$

But the astronomical second is an arbitrary unit. A refined value of s/τ gives $\tau = 0.0108837319$ s, or $91.880204932 = \tau/s$. It is this (planetary) value of τ in seconds that corresponds to the relation mm/$\lambda = (2/g - 1)^{1/8} = 0.9422119204$, or 1 meter $= 1061.332359$ λ. Adjustment to these two metric equivalents, both having the form $f(\pi, 2)$,[9] is most conveniently supplied by the semimajor axis and the sidereal period of Mercury (though Venus or Earth would serve.)

Planetary elements in GU are tabulated below (see table 2).[10] The final column shows the calculation of $\beta(x')$ for each sidereal period; the use of this function will be illustrated by the first three planets in non-Keplerian relations.

The internal consistency of the formulations above, and their utility in the detection of non-Keplerian planetary relations, is established for the innermost three planets by consideration of their respective values for $\beta(x')/10^7$:

x' $10^{-7}\beta(x')$ Non-Keplerian relations

m' $10^{-7}m$ $m = 0.6160888767 \times 10^{14}$
                    Measured period, 87.969256 Earth days
                    Calculated in τ, $69.83398518 \times 10^7\tau$

---

8   The concept of squaring can be defined as division by the reciprocal; when dealing with ratios, which are not literally magnitudes, that is clearer than the arithmetical definition.

9   It may seem that $f(\pi, 2, 10)$ should be written, but in this matter 10 operates only as the point-shift symbol, as it does in any place-system of numeration. It has nothing to do with the *number* ten, but indirectly with binary 10, which stands for two.

10   All semimajor axes and sidereal periods in GU were obtained from those of Mercury by applying the measured ratios reported in readily available astronomical tables.

Table 2 (in GU)*

$$m = (2\pi\sqrt{m'})^4/10^7g \qquad m' = \sqrt{(10^7gm)}/(2\pi)^2$$

| Semimajor axis | | Sidereal Period | | $(2\pi\sqrt{x'})^4/10^{14}g$ |
| GU | | | GU | $= \beta(x')$ |
$\lambda \times 10^{14}$		Earth days	$\tau \times 10^7$	$\tau \times 10^7$
m 0.61608887	m'	87.969256	m' 69.83398468	10^7m (in λ)
V 1.15122102	V'	224.700789	V' 178.3776761	4.019673
E 1.59155499	E'	365.25636	E' 289.9570624†	10.621286
M 2.42503877	M'	686.979702	M' 545.3556409	37.572434
A 4.44521082	A'‡	1708.89134	A' 1356.16326	232.34518
J 8.28054866	J'	4332.5878	J' 3439.4048	1494.43249

* This table was originally formed by using the expression for s/τ first shown, before that was perceived to be not quite exact, thanks to the traditional fractions 1/24 and 1/60 surviving in our second from Babylonian astrological tradition. Division of time without arbitrary fractions is reflected in GU. Repeated doublings or halvings suffice, though they are obscured by decimal fractions.

† $(10^{14}/m)^{64} = 289.98377 \times 10^{11}$ [τ] $\cong E' \times 10^4$

‡ For the asteroids, calculation of a period was made according to Kepler's law. If there were a single planet at this distance, its sidereal period would be A'; $10(g - 1) = 233.7 \cong \beta(A')$.

In this calculation the semimajor axis of Mercury in GU is obtained from its sidereal period by the β-function, refining the value of an astronomical second in terms of τ: $69.83398518 \times 10^7\tau = 87.969256 \times 24 \times 3600 \times \tau/s$, 1 second $= 91.88024932 \ \tau$, or τ is the duration of 0.0108837319 astronomical second of time.

v' 4.019673 $\beta(V')/(10^7m) = (V'/m')^2$
Implied $(V'/m') \times m'$ yields V'
at 224.7007906
Measured period in Earth days
$= 224.7007890$

This calculation serves to assure internal consistency in use of the β-function to relate the sidereal period of Venus in τ to Mercury's semimajor axis in λ. That is a non-Keplerian relation, in the sense that it does not require use of the semimajor axis of Venus or of the exponent 3/2. The ratio of sidereal periods V'/m' is found from the root-extraction made after shift of the decimal point – a shift necessitated by the move from 10^{14} λ to 10^7 τ in Galilean units.

E' 10.621286 $V' \times \sqrt{[\beta(E')/\beta(V')]}$ yields $E' = 365.25636$

Measured period, Earth days   365.25636

Calculated as $V'(m'^{1/4}/10^2)$ 365.24704

In this final illustrative calculation, a quite unanticipated non-Keplerian relation emerges – that of the 4th root of Mercury's sidereal period to the period of Earth (with two point-shifts), involving the sidereal period of Venus in a way that suggests its focal value to be very nearly related to the Earth's tropical year.

The speeds of planets vary continuously in their elliptical orbits. Using the reported eccentricities from table 1, the radii of the circles equal in area to each of the first three orbits in GU can be calculated, and the implied orbital circumferences. The speed of the planet can then be determined in λ/τ as shown in table 3.

The first circumference yields $(m\odot/2)^4 = 13.44041036 \cong (J/m)$ $\cong (2V/m - 2)(2V/m + 4) = 13.44098276$; $J/m = 13.44051011$ at present.

The circular orbit for Venus in table 3, $V\odot$, is $10V/E$, which is the semimajor axis of Venus as shown in ordinary tables in AU but with the decimal point shifted one place to the right. That non-Keplerian relation of Venus to Earth could hardly have been anticipated, and it throws interesting new light on Bode's law. Under that empirical rule, as first stated in 1772, the distance unit was AU/10. Announced before the discovery of Uranus, it gave rough distances of planets from the Sun (to 2 significant figures), and included the asteroid belt, but failed at Neptune.[11] The rule was that numbering the planets outward from the Sun, each distance is $4 + 3 \times 2^{(n-2)}$; for Earth, $n = 3$ and its distance would be 10 in Bodeian units. If instead of the Bodeian $E = 10$ we put $E = 10/2\pi$, all the semimajor axes are very nearly in GU, with Mercury's closely approximating $g/2$.

Thus even at the present advanced state of science there are still some potential uses of Galilean gravitational units that may

---

11 See my *History of Free Fall* (Toronto 1989) for more on Bode's law and its relation to GU.

Table 3 (in GU)
Speeds in circles of equal area to orbital ellipses

Sidereal period $\times 10^7 \tau$		Radius $\times 10^{14} \tau$		Circumference $\times 10^{14} \lambda$		Speed λ/τ	
m'	69.8339853	$m^\dagger$	0.6094713	$m\odot$	3.829421117	v_m	54,836
V'	178.3776761	$V^\dagger$	1.1512012	$V\odot$	7.233210465	v_V	40,540
E'	289.957064	$E^\dagger$	1.5897382	$E\odot$	9.983816959	v_E	34,432

repay exploration by astronomers. The pioneer of modern physics discovered more things in heaven and earth than had been dreamt of by philosophers. Those who are skilled in astronomical theory and who possess powerful calculational facilities can readily determine whether, and to what extent, what has been said above is merely coincidentally in agreement with precise measurements, and how much of it, if any, can be put to modern theoretical use.

Bibliography

The ensuing listing is divided into categories of ancient, medieval, sixteenth-century, and seventeenth-century science, following general works, biographies of Galileo, and English translations of his scientific writings.

General Works

Butts, R.E., and Pitt, J.C., eds. *New Perspectives on Galileo*. Dordrecht 1978

Clavelin, M. *The Natural Philosophy of Galileo*. Cambridge (Mass.) 1974; first French edition 1968

Cohen, I.B. *The Newtonian Revolution*. Cambridge 1980

Dijksterhuis, E.J. *The Mechanization of the World Picture*. Oxford 1961; first Dutch edition 1950

Finocchiaro, Maurice. *The Galileo Affair: A Documentary History*. Berkeley 1989

Hall, A.R. *The Revolution in Science, 1500–1750*. London 1983

Holton, G. *Introduction to Concepts and Theories in Physical Science*. Reading (Mass.) 1958

Koyré, A. *Galileo* [sic. for *Galilean*] *Studies*. Hassocks 1978; first French edition 1939

– *Newtonian Studies*. Cambridge 1955

– *Metaphysics and Measurement*. Ed. M.A. Hoskin. London 1968

Kuhn, T. *The Copernican Revolution*. New York 1959

– *The Structure of Scientific Revolutions*. Chicago 1962

McMullin, E., ed. *Galileo Man of Science*. New York 1967

Shea, W.R. *Galileo's Intellectual Revolution*. London 1972

– ed. *Nature Mathematicized*. Dordrecht 1983

Wallace, W.A. *Prelude to Galileo*. Dordrecht 1981

– *Galileo and His Sources*. Princeton 1984

Biographies of Galileo

Drake, S. *Galileo at Work*, Chicago 1978
Drinkwater[-Bethune], J.E. *The Life of Galileo*. London 1829
Fahie, J.J. *Galileo: His Life and Work*. London 1903
Geymonat, Ludovico. *Galileo Galilei*. Trans. S. Drake. New York 1965; first Italian edition 1957
Paschini, P. *Vita e Opere di Galileo Galilei*. Rome 1965
Viviani, V. *Racconto istorico della vita di Galileo* [1657]. Ed. S. Salvini. Florence 1717
Wohlwill, Emil. *Galilei und sein Kampf ...*, vol. 1. Hamburg 1909

Scientific Writings of Galileo in Recent English Translation

1584. *Universe* in W.A. Wallace, *Galileo's Early Notebooks: The Physical Questions*. Notre Dame 1977
1586. *The Sensitive Balance* in L. Fermi, *Galileo and the Scientific Revolution*. New York 1961
1587. *Problems of Motion* (dialogue) in Drake and Drabkin, *Mechanics in Sixteenth-Century Italy* (see under 'Renaissance Science,' below)
1587–90, *Memoranda on Motion* in Drake and Drabkin, *Mechanics*
1588. *Change and Elements* in Wallace, *Galileo's Early Notebooks* (see '1584,' above)
1590. *On Motion* in I.E. Drabkin and S. Drake, *Galileo on Motion and on Mechanics*. Madison 1960
1601. *Mechanics* in Drabkin and Drake, *Galileo on Motion*
1602. *Letter to Guidobaldo del Monte* in Drake, *Galileo at Work*, 69–71
1602–37. *Notes on Motion* in S. Drake, monograph 3, *Annali dell' Istituto e Museo di Storia della Scienza*. Florence 1979
1605. *Dialogue on the New Star* in S. Drake, *Galileo against the Philosophers*, Los Angeles 1976
1607. *Operations of the Geometric and Military Compass*. Washington: 1978
1610. *Sidereus Nuncius or The Sidereal Messenger*. Trans. A. Van Helden. Chicago 1989
– *The Starry Messenger* in S. Drake, *Telescopes, Tides, and Tactics*. Chicago 1983. Abridged in *Discoveries* (see '1613,' below)
1612. *Bodies That Stay atop Water, or Move in It* in S. Drake, *Cause, Experiment, and Science*. Chicago 1981
1613. *Letters on Sunspots* in S. Drake, *Discoveries and Opinions of Galileo*, New York 1957 (abridged text)

– *Letter to Castelli* in Drake, *Galileo at Work*, 224–9

1615. *Old and New ... Doctrines of Holy Scripture in Purely Physical Conclusions* [*Letter to Christina*] in Drake, *Discoveries*

1619. *Discourse on the Comets* (ed. M. Guiducci) in S. Drake and C.D. O'Malley, *Controversy on the Comets of 1618*, Philadelphia 1960

1623. *The Assayer* in Drake and O'Malley, *Controversy*; abridged text in Drake, *Discoveries*

1624. *Reply to Ingoli* in Finocchiaro, *The Galileo Affair*

1631. *Opinion on Flood Control on the Bisenzio River* in Drake, *Galileo at Work*, 321–9 (abridged text)

1632. *Dialogue concerning the Two Chief World Systems*. Trans. S. Drake. Berkeley and Los Angeles 1953, 1962, 1967

– *Dialogue on the Great World Systems*. Trans. T. Salusbury; ed. G. de Santillana. Chicago 1953

1638. *Dialogues concerning Two New Sciences*. Trans. H. Crew and A. de Salvio. New York 1914; 2d ed., Evanston and Chicago 1939

– *Two New Sciences*. Trans. S. Drake. Madison 1974; 2d ed., Toronto 1989

1641. *On Euclid's Definitions of Ratios* in Drake, *Galileo at Work*, 422–36

Ancient Greek Science

Archimedes. *Works*. Ed. T.L. Heath, Cambridge 1897

Clagett, M. *Greek Science in Antiquity*. New York 1955

Cohen, M.R., and Drabkin, I.E. *A Source Book in Greek Science*. New York 1948

Drake, S. *History of Free Fall*. Toronto 1989

– 'Hipparchus – Geminus – Galileo,' *Studies in History and Philosophy of Science* 20 (1989), 47–56

– 'Ptolemy, Galileo, and scientific method,' *Studies in History and Philosophy of Science* 9 (1978), 99–115

Heath, T.L. *Aristarchus of Samos*, Oxford 1913

– *Mathematics in Aristotle*. Oxford 1949

Hero of Alexandria. *The Pneumatics*. Ed. M.B. Hall. London 1971

Pappus of Alexandria. *La Collection Mathématique*. Ed. P. Ver Eecke. Paris – Bruges 1933

Ptolemy, *The Almagest* (trans. R.C. Taliaferro) in vol. 16 of *Great Books of the Western World*. Chicago 1938

Medieval Science

Clagett, M. *The Science of Mechanics in the Middle Ages*. Madison 1959
- *Archimedes in the Middle Ages*. Philadelphia 1976–83
- ed. *Nicole Oresme and the Medieval Geometry of Qualities and Motions*. Madison 1968
Crombie, A.C. *Medieval and Early Modern Science*. Cambridge (Mass.) 1961
Crosby, H.L. *Thomas of Bradwardine his Tractatus de Proprotionibus* ... Madison 1955
Drake, S. 'Bradwardine's function, mediate denomination, and multiple continua,' *Physis* 11 (1970), 51–68
- 'Euclid Book v. From Eudoxus to Dedekind' in I. Grattan-Guinness, *History in Mathematics Education*. Paris 1987
- 'Free fall from Albert of Saxony to Honoré Fabri,' *Studies in History and Philosophy of Science* 5 (1975), 347–66
- 'A further reappraisal of impetus theory,' *Studies in History and Philosophy of Science* 7 (1976), 319–36
- 'Impetus theory and quanta of speed ...,' *Physis* 16 (1974), 47–65
- 'Impetus theory reappraised,' *Journal of the History of Ideas* 36 (1975), 27–46
- 'Medieval ratio theory ...,' *Isis* 64 (1973), 67–77
Lindberg, D.A., ed. *Science in the Middle Ages*. Chicago 1978
Maier, A. *Die Vorläufer Galileis im 14. Jahrhundert*. Rome 1949
Moody, E.A., and Clagett, M. *The Medieval Science of Weights*. Madison 1952
Oresme, N. *De proportionibus proportionum: Tractatus latitudinibus formarum*. Ed. B. Politi. Venice 1505
- *Le livre du ciel et du monde*. Trans. A.D. Menut and A.J. Denomy. Madison 1968
- *Tractatus de commensurabilitate in motuum coelorum*. Trans. E. Grant. Madison 1971
Wilson, C. *William Heytesbury: Medieval Logic and the Rise of Mathematical Physics*. Madison 1956

Renaissance Science

Benedetti, G.B. Translations in Drake and Drabkin, below
Commandino, F. *Liber de centro gravitatis*. Bologna 1565
del Monte, G. Translations in Drake and Drabkin, below

Drabkin, I.E., and Drake, S. *Galileo on Motion and on Mechanics*. Madison 1960

Drake, S. 'Early science and the printed book,' *Renaissance and Reformation* 6 (1970), 43–52

– 'The evolution of *De motu*,' *Isis* 67 (1976), 239–50

– 'Galileo and the first mechanical computing device,' *Scientific American* 234 (1976), 104–13

– 'Galileo's pre-Paduan writings: Years, sources, motivations,' *Studies in History and Philosophy of Science* 17 (1986), 429–48

– 'Origin and fate of Galileo's theory of tides,' *Physis* 4 (1962), 185–94

– 'Renaissance music and experimental science,' *Journal of the History of Ideas* 31 (1970), 483–500

– 'Tartaglia's *squadra* and Galileo's *compasso*,' *Annali dell' Istituto e Museo di Storia della Scienza di Firenze* 2 (1977), 35–54

Drake, S., and Drabkin, I.E. *Mechanics in Sixteenth-Century Italy*. Madison 1969 (texts of Tartaglia, Benedetti, del Monte, Galileo)

Dreyer, J.L.E. *Tycho Brahe*. Edinburgh 1890

Edwards, W.F., and Wallace, W.A. *Galileo's* Tractatio de praecognitionibus *and* Tractatio de demonstratione. Padua 1989

Koyré, A., ed. *La science au seizième siècle*, Paris 1960

Rose, P.L., and Drake, S. 'The pseudo-Aristotelian *Questions of Mechanics*,' *Studies in the Renaissance* 18 (1971), 65–104

Stevin, S. *The Principal Works*, vol. 1 (ed. E.J. Dijksterhuis). Amsterdam 1955

Swerdlow, N.M. and Neugebauer, O. *Mathematical Astronomy in Copernicus's De revolutionibus*. New York 1984

Taisnier, J. *Opusculum* ... Cologne 1562

Tartaglia, N. *Regola generale da sulevare ... ogni affondata nave*. Venice 1551; other works trans. in Drake and Drabkin, above

Varro, M. *Tractatus de motu*. Geneva 1584

Wallace, W.A., 'The enigma of Domingo de Soto,' *Isis* 59 (1968), 384–401

Seventeenth-Century Science

Aiton, E.J., 'Galileo's theory of the tides,' *Annals of Science* 10 (1954), 44–57

Baliani, G.B. *De motu naturali gravium* ... Genoa 1646

Beeckman, I. *Journal* ... (ed. C. De Waard). The Hague 1939–53

Bell, A.E. *Christian Huygens and the Development of Science*. London 1947

Boas, M. *The Scientific Renaissance*. New York 1962

Caspar, M. *Kepler* (trans. C.D. Hellman). New York 1959; first German edition 1948

Cazré, P. *Physica demonstrativa* ... Paris 1645

Cohen, I.B. *The Birth of a New Physics*. New York 1985

– ed. *Isaac Newton's Papers and Letters on Natural Philosophy*. Cambridge (Mass.) 1958

Costabel, P. *Leibniz and Dynamics*. London 1973; first French edition 1960

Drake, S. *Discoveries and Opinions of Galileo*. New York 1957

– *Galileo against the Philosophers*. Los Angeles 1976

– *Galileo Studies*. Ann Arbor 1970

– 'Alleged departures from Galileo's law of fall,' *Annals of Science* 38 (1981), 339–42

– 'Analysis of Galileo's experimental data,' *Annals of Science* 39 (1982), 389–97

– 'Galileo and mathematical physics,' in *Scienza e Filosofia*, ed. C. Mangione, Milan 1985

– 'Galileo and satellite prediction,' *Journal for the History of Astronomy* 10 (1979), 75–95

– 'Galileo, Kepler, and phases of Venus,' *Journal for the History of Astronomy* 15 (1984), 198–208

– 'Galileo's accuracy in measuring horizontal projections,' *Annali dell' Istituto e Museo di Storia della Scienza di Firenze* 10 (1985), 1–14

– 'Galileo's constant,' *Nuncius* 2 (1987), 41–54

– 'Galileo's experimental confirmation of horizontal inertia,' *Isis* 64 (1973), 291–305

– 'Galileo's first telescopic observations,' *Journal for the History of Astronomy* 7 (1976), 153–68

– 'Galileo's physical measurements,' *American Journal of Physics* 54 (1986), 302–6

– 'Galileo's "Platonic" cosmology and Kepler's *Prodromus*,' *Journal for the History of Astronomy* 4 (1973), 174–91

– 'Galileo's steps to full Copernicanism, and back,' *Studies in History and Philosophy of Science* 18 (1987), 93–105

– 'Mathematics and discovery in Galileo's physics,' *Historia Mathematica* (1974), 129–50

– 'Measurements in Galileo's science,' *History of Technology* 5 (1980), 39–54

– 'The role of music in Galileo's experiments,' *Scientific American* 232 (June 1975), 98–104

– 'The tower argument in the *Dialogue*,' *Annals of Science* 45 (1988), 295–302

– 'Velocity and Eudoxian proportion theory,' *Physis* 15 (1973), 49–64

Drake, S., and Kowal, C.T. 'Galileo's sighting of Neptune,' *Scientific American* 243 (December 1980), 74–81

Drake, S., and MacLachlan, J. 'Galileo's discovery of the parabolic trajectory,' *Scientific American* 232 (March 1975), 102–10

Fabri, H. *Tractatus physicus de motu locali*. Lyons 1646

– *Dialogi physici*. Lyons 1665

Gassendi, P. *De proportione qua gravia decidentia accelerantur*. Paris 1646

Grassi, O. [= Sarsi, L.]. Translations in Drake and O'Malley, *Controversy on the Comets of 1618*. 1960

Hall, A.R. *Ballistics in the 17th Century*, Cambridge 1952

Hill, D.K. 'Dissecting trajectories,' *Isis* 79 (1988), 646–68

– 'Galileo's work on f 116v: A new analysis,' *Isis* 77 (1986), 283–91

Huygens, C. *The Pendulum Clock*. trans. R.J. Blackwell. Ames (Iowa) 1986; original Latin edition Paris 1673

– *Treatise on Light*. Trans. S.P. Thompson. Chicago 1950; original French edition Leiden 1690

Kepler, J. *The Secret of the Universe*. Trans. A.M. Duncan. New York 1981; original Latin editions 1596, 1621

Kowal, C.T., and Drake, S. 'Galileo's observation of Neptune,' *Nature* 287 (1980), 277–8

Koyré, A. 'Documentary history of the problem of fall,' *transactions of the American Philosophical Society*, n.s. 45, pt. 4 (1955)

– *The Astronomical Revolution*, Ithaca 1973; first French edition Paris 1961

Le Tenneur, J.A. *De motu naturaliter accelerato*. Paris 1649

MacLachlan, J. 'Galileo's experiments with pendulums: Real and imaginary,' *Annals of Science* 33 (1976), 173–85

– 'A test of an "imaginary" experiment of Galileo's,' *Isis* 64 (1973), 374–9

More, L.T. *Isaac Newton*. London 1934

Moscovici, S. *L'expérience du mouvement*. Paris 1967

Naylor, R. 'Galileo and the problem of free fall,' *British Journal of the History of Science* 7 (1974), 105–34

– 'The role of experiment in Galileo's early work on the law of fall,' *Annals of Science* 37 (1980), 363–87

– 'The search for the parabolic trajectory,' *Annals of Science* 33 (1976), 153–74

Newton, I., *Mathematical Principles of Natural Philosophy*. Ed. F. Cajori. Berkeley 1947; first Latin edition London 1687

Settle, T. 'An experiment in the history of science,' *Science* 133 (1961), 19–23

Van Helden, A. 'The invention of the telescope,' *Transactions of the American Philosophical Society* 67, pt. 4 (1977)

Wallace, W.A., ed. *Reinterpreting Galileo*. Washington 1986

Westfall, R.S. *Force in Newton's Physics*. London 1971

Wisan, W. 'The new science of motion: A Study of Galileo's *De motu locali*,' *Archive for the History of Exact Sciences* 13:2/3 (1974), 103–306

Index